[美国] 亚历克斯·罗兰 著　付满 译

战争与技术

牛津通识读本·

War and Technology

A Very Short Introduction

译林出版社

图书在版编目（CIP）数据

战争与技术 /（美）亚历克斯·罗兰（Alex Roland）著；付满译. —南京：译林出版社，2023.7
（牛津通识读本）
书名原文：War and Technology: A Very Short Introduction
ISBN 978-7-5447-9676-7

Ⅰ.①战… Ⅱ.①亚… ②付… Ⅲ.①军事技术－技术史－世界 Ⅳ.①E9-091

中国国家版本馆 CIP 数据核字（2023）第 069568 号

著作权合同登记号 图字：10-2018-429号

战争与技术 ［美国］亚历克斯·罗兰 / 著 付 满 / 译

责任编辑 陈 锐
装帧设计 景秋萍
校 对 孙玉兰
责任印制 董 虎

原文出版 Oxford University Press, 2016
出版发行 译林出版社
地 址 南京市湖南路 1 号 A 楼
邮 箱 yilin@yilin.com
网 址 www.yilin.com
市场热线 025-86633278
排 版 南京展望文化发展有限公司
印 刷 江苏扬中印刷有限公司
开 本 890 毫米 ×1260 毫米 1/32
印 张 8.625
插 页 4
版 次 2023 年 7 月第 1 版
印 次 2023 年 7 月第 1 次印刷
书 号 ISBN 978-7-5447-9676-7
定 价 39.00 元

序　言

陈仲丹

对“牛津通识读本”我早就知晓。这是一套邀请专家写的通识（不是通俗）读物。该丛书由牛津大学出版社组织编写，“大家写小书”，篇幅不大，所涉范围广泛，是以导读形式将各门学科内容介绍给大众的书系。译林出版社引进出版了该丛书的简体中文版，且在书中保留英语原文，以便读者在看完汉译后还能回溯品味原著的文字。前些年，译林出版社到我工作的南京大学介绍该丛书，适与当时学校大力提倡的通识教育相合。现在我与该丛书的因缘又近了一层，有幸为美国学者亚历克斯·罗兰所著的《战争与技术》作序推荐。

在此，我要先对自己与军事史的关系有所交代。一者，我是在大学中为数不多有预备役军官身份的教师，军衔为上校，还得到作训服和常服各一套，身穿军服参加过部队的活动。二者，我在南京大学开设了有关兵器史的课程，并拍摄了课程视频，供国内的大学选用。在课程中，我将几千年的兵器史分为冷兵器、黑火药兵器、近代兵器、现代兵器几个阶段。三者，我编写过几本与战争有关的书，如《图说兵器战争史：从刀矛到核弹》《太平洋战场》等。我现在仍是国防部和南京市的国防教育专家库成员，自然关注"战争和技术"的话题。

此次有作序的近便，我得以先睹为快，成为《战争与技术》中文书稿最早的一批读者。在开卷之后，我突然想起曾与本书的作者在多年前就有交往。1997年，我受学校派遣去美国杜克大学的亚太研究中心访学，不久就转到该校的历史系交流。在杜克大学历史系接待我的就是本书作者、历史系主任亚历克斯·罗兰。他个子不高，办事干练，一见面就让我使用一位在英国访学的教师的办公室，并为我写信介绍给有关老师去听课。据罗兰先生自我介绍，他的专业是军事史，本科毕业于美国安纳波利斯海军学院，学的是海军工程，研究生阶段在历史系就读，取得博士学位，他是在军事学的大背景下由工学转向史学。后来我在图书馆看了他上课的录像，讲的是军事上的防御，从堡垒、防御工事讲到第二次世界大战中使用的雷达，最后谈到美国的"星球大战计划"（反弹道导弹防御系统），整个课程以防御为线索贯通古今。罗兰先生现已退休，是杜克大学历史系的荣休教授，最近他还就美国私营企业从事航天事业发表看法。我回国后对军事史有了兴趣，开设课程，写军事史的书，说来与罗兰先生有直接的关联。

《战争与技术》一书是亚历克斯·罗兰以他2014年在美国西点军校讲授的同名课程发展而来的。西点军校是美国最著名的军事院校，其地位与俄罗斯的伏龙芝军事学院以及中国的国防大学相当。西点军校的老师历来有着"将军之师"的称誉，能被邀请到该校授课，足以见得罗兰先生的学术地位。

该书的核心主旨，正如作者在开篇所言："本书的目的，是想从最早期的人类到现在的经历当中，追踪技术和作战的共同演进过程。""技术，是为追求人类目标而试图改变物质世界；作战，则通过威胁和动武来试图改变人类行为。"因而有必要了解这两者之间相互联系的全过程。一般而言，任何时代最先进的技术成果都会首先服务于战争的需要，"国之大事，在祀与戎"，战争关乎国家兴亡和百姓福祉，自然有着不容忽视的优先地位。在战争中，军事技术和其他技术通常会有迅速发展的机会。在同属"牛津通识读本"丛书的《第二次世界大战》中有这样的记述："新式坦克、飞机、战舰以及火炮得以开发并投入实战，雷达、喷气式飞机、弹道导弹、核武器是战时投入使用的最了不起的技术成果，战后得到持续发展。""大规模输血以及新型药物（如青霉素）挽救了千万伤兵的性命，并成为战后的基础医疗方法。所以说，虽然'二战'期间物质破坏巨大，但毕竟也有几样有益的技术进步。"新中国建立后举全国之力从事的"两弹一星"事业，也主要是让在第二次世界大战中开发的技术成果成为保卫家园的利器。

在中国近代史上有一句熟语"落后就要挨打"，而落后主要指的是技术落后，在对手"船坚炮利"的优势面前处于弱势。因而，本书作者特别强调技术的物质性，认为这是"人类对物质世界的一种有目的的操控"。当然，我们也要看到，技术的物质优势是相对的，作为硬件的技术优势固然重要，而作为软件的精神力量也不容忽视，不然就无法解释在中国革命战争中，装备落后的人民军队为何能屡屡战胜武器精良的敌人。晚清时期，湘军统帅曾国藩收到李鸿章的来信，李在信中盛赞上海洋人的火器

"神乎技矣"。曾国藩的回答是:"真美人不甚争珠翠,真书家不甚争笔墨;然则将士之真善战者,岂必争洋枪洋药乎!"曾的看法也有其道理,不然就无法解释,后来的淮军在物质的技术条件已与对手相当时,却不能在甲午战争中战胜对手。

本书是一本篇幅不大的导论性著作,当然不能对战争与技术的关系做全面探究。即便如此,该书的写法也会不时给读者以惊喜,其详略有其独到的取舍。比如在介绍早期武器时,本书重点介绍了三十万年前海德堡人使用的"舍宁根标枪",还配了标枪图片,说明这是迄今为止世界上发现的最早的武器,且可作军民两用,用于狩猎和作战。对于这一古标枪,通常的军事史著作中很少提及。有意思的是,作者还刻意说明这种标枪是非对称作战的武器,用于"打了就跑"的战术。而21世纪用于袭击的简易爆炸装置,在他看来是新时代的"舍宁根标枪"。这样的比照使古今历史有了联系,使得对历史的描述有了纵深感。

另外,本书的写法还有从大技术角度着眼的特点。作者认为,用于战争的技术不应局限于直接使用的武器,还可包括其他方面。罗兰先生擅长的对防御的研究在本书中也有体现。他重点介绍了巴勒斯坦六千年前的防御城堡杰里科古城,并配了图片。而他关注的军事上的大技术还包括道路修建,罗马大道便于在帝国境内迅速调动军队,这也是军民两用的技术产品。他还认为,火箭是更高层次的军民两用技术,将战争从陆地延伸到海洋、天空后又扩展到太空。

正是作者致力于从宏观的视野来考虑战争与技术的关系,才使得本书与许多同类著作有所不同。他不太在意介绍具体的知识,而是侧重于关注现象、事件之间的联系,提出了一系列富有深意的命题，如对称作战和非对称作战，不同兵种的联合作战范式，火药革命宣告“化学能时代”或“碳时代”的到来，以及区别于“技术决定论”的“技术的动能”，“军工联合体”的产生，等等。这样的写法使得本书带有浓烈的问题意识，有着区别于一般知识读物的理论色彩，体现出作者的哲思。我建议读者在阅读本书时可与知识类的军事史著作交互阅读，以求相互补充，以便在熟悉历史背景和具体细节后更好地理解本书的内涵。

本序作为对导读类著作的导读，最后要说明的是，源于教育背景和知识结构的特点，罗兰先生更为熟悉西方的文化语境，因而书中所用例证多来自欧美国家，对东方尤其是中国的历史资源引用不多。因而，读者在阅读本书时也应蓄积一些与本国相关的知识储备，这样立足本土、放眼全球，视野就会更开阔。

目 录

致 谢 1

第一章 前 言 1

第二章 陆 战 7

第三章 海战、空战、太空战和近现代作战 40

第四章 技术变革 80

结 语 108

术语表 113

索 引 117

英文原文 131

致 谢

本书第一稿成形于2014年春季学期，当时我在美国西点军校教授“技术与战争”这门历史课。参与这门课学习的学员有：泰勒·艾伦、麦肯齐·比斯利、乔纳森·克鲁西提、摩根·丹尼森、罗伯特·菲、雅格·方丹、卢卡斯·霍奇、布莱恩·霍普、亚历克斯·库卡斯基、詹姆斯·奥基夫、亚历山大·里弗斯、特拉维斯·史密斯和道格拉斯·泰勒。感谢他们所有人在这个专题上帮助我磨砺思想、启发实效，他们的每次质疑都让我的讲课变得更为清晰和更加有趣。本书或许难以实现满足所有人的期望，但学员们的贡献足以让它生色良多。

本书成稿后，我请了四位好朋友丹尼尔·海德里克、韦恩·李、马修·莫顿和埃弗雷特·惠勒对它提出批评意见，他们每一位都是知名学者。在认真仔细地阅读了原稿后，他们提出了不少有益的修改建议。这些建议帮我减少了书中的错漏，避免书稿仓促付梓可能会带来的尴尬。我从韦恩·李那里获益尤多，因为他的皇皇巨著《发动战争：世界历史中的冲突、文化和创新》（牛津大学出版社2016年版）涵盖的是同一领域，且更为iii 详尽。牛津大学出版社延请的三位圈外读者也对原稿提出了建设性的批评意见，我一并接纳并在书中有所体现。

牛津大学出版社的南希·托夫身体力行，证明了一位敬业、勤奋的编辑的支持有多么重要！她能干的助手埃尔达·格拉纳塔，堪称效率、热心和文雅的典范。文字编辑本·萨多克也是一个精明、温和、能干且热心的人。

我的妻子莉兹按照我的严苛要求做了本书的索引。她工作迅捷、准确，思虑缜密且有着不懈的热情，是我最好的朋友、搭档、批评者和支持者。

书中如果依然存在错误和不足，那么责任在我，与上述人 iv 士无关。

第一章

前　言

人类是带着武装来到世间的。早在智人开始在大地上行走之前，早期原始人就已经为特定的目的打造和使用武器了。这些武器肯定是在狩猎中运用，也可能被用于作战。制造和使用武器，以及其他用于军事的技术，正是人类成其为人的一部分。本书的目的，是想从最早期的人类到现在的经历当中，追踪技术和作战的共同演进过程。

技术和作战，本质上都是物质的。它们是为了实现人类目的而改变物理世界的共有进程。技术，是为追求人类目标而试图改变物质世界；作战，则通过威胁和动武来试图改变人类行为。这两个现象在物理和物质上紧密相连。本书的第二个目的，就是追踪这种紧密联系的演进过程。

全书有一个贯彻始终的主题思想，即，技术是改变作战的最大变量。政治、经济、意识形态、文化、战略、战术、领导力、哲学、心理学等各种因素都影响作战，但其中没有一种变量能像技术这样，完整地解释了史前作战到现代作战的发展过程。从石器

1 时代到核能时代，技术推动了作战的演进。

一个简略的思想实验有助于我们理解这个结论。想象一下：如果亚历山大大帝穿越到21世纪的第二个十年，被派去征服阿富汗，他能完成这项任务吗？他曾经在公元前330年征服过这片地域。在一场长达十三年的征战中，他从马其顿的家乡一路征伐，打到了今天的希腊、土耳其、叙利亚、黎凡特、埃及、伊拉克、巴基斯坦、阿富汗，甚至更远的地方。他击败了当时最强的军队，穿越沙漠和山地，带着各种无法从当地购买或劫掠的物资，也在身后留下相对的和平与稳定。这番征战证明了他是有史以来最伟大的统帅之一，一位深谙战争艺术的公认大师。

他理解并出色地运用了作战学员们所谓的"战争原则"。这些原则在表述上各有不同，但大体类似编入《美军战地手册3-0》（2011年版）里的九条原则：保持客观、主动进攻、集合作战、节省兵力、善于机动、统一指挥、注意安全、出其不意和化繁为简。这些原则与其说是实际作战中要执行的守则，不如说是对战争进行分析归纳而总结出的一份清单。但是，专家们一直把它看成作战取胜的关键。安托万-亨利·约米尼男爵（拿破仑的部下和学生）曾说过："无论使用什么样的武器，也无论处于什么样的时空，战争的原则都独立存在、不会改变。"如果亚历山大大帝在公元前4世纪就掌握了这些战争原则，他一定会在21世纪同样有效地运用它们。这些原则不会告诉他如何思考，但会告诉他该考虑些什么。毋庸置疑，他在古代作战中所看重的这些原则，换到今天的战场依然会被同等重视。

但技术却带来了疑问。这位穿越重生的亚历山大大帝不懂

2 也无法学习的一样东西，就是技术。他如何理解炸药、飞机、卫

星、无线电、电脑或精确制导武器？生活在发达国家的现代人对这些技术习以为常，多少了解一些飞机和直升机怎么能在天上飞，卫星为什么能在轨道运行，东西爆炸是如何发生的，电磁频谱里藏着哪些性能等问题涉及的知识。那么，没等亚历山大大帝弄明白这些“奇迹”如何发生的时候，他在阿富汗的战事应该就已经结束了。现代战争的其他方面，他都了解或可以学习；唯独技术，使得现代作战迥异于他平生所了解的作战，让他无法理解。正如约米尼所领悟到的：作战的根本法则，超越时间，亘古不变。而技术，却是在不断变革，并在此过程中改变着作战。它是作战变革的主要推动力。正是这个变量，使得亚历山大大帝在今天会显得无能为力。本书想要揭示的问题就是：在人类历史进程中，那些变革如何展现出特定的形式，以及它们背后的成因。

书中的章节安排按照个人的想法编排，希望有助于讲清问题。首先，笔墨着重于先前。也就是说，前现代的作战是本书关注的重点。一些观念在久远的过去，就已在人类的行为实践中扎根。这些罗列在本书“术语”部分的观念，为我们理解令人眼花缭乱的现代战争技术提供了一把钥匙——这也是本书的一个前提。其次，本书聚焦于不断变化的军事技术所带来的一个非常突出却看似矛盾的结果——也是本书的次级主题。纵观历史，先进技术一般有助于获胜，但并不能提供胜利保证。使用更新更好军事技术的一方，不见得一定就是赢家。作战所运用到的技术，总体而言并非绝对有效。它的价值，其实只与敌人的能力有关。把作战设想成一场决斗，其中任何一方都可以选择自己的武器。而对任何一方而言，武器的选择就会影响到交战规则（包括无交战规则）、战略、战术、政治、外交、环境以及其他战斗条件。比如，一方选择手枪，另一方选择用剑，那么决斗的结果其实早已注定。但如果另一方选择的是步枪而不是剑，结果很可能又会逆转。虽然使用手枪这一技术条件没有发生变化，但是它的相对有效性却败于对手。

本书还提出一点：自古到今，技术和作战一直相互作用。作战对技术的改变程度差不多等同于技术对作战的改变。我们将略微简化惯常的时代划分，按照时间顺序来探究这两者之间的辩证关系。从史前作战开始，经新石器时代、远古、古典时期、中世纪、近代早期一直到现代。跨越这些基本的时代分割的，是那些运用到独特军事技术的时期。我们将追溯军事技术中的能量驱动方式——从体能、风能到碳基化学能，再到核能。战争在物理空间展开的方式，也有自己的时间顺序。陆战这种最古老也最复杂的战争形式，历史最为悠久。关于它的故事，本书将按传统的时代划分分段论述，并在重点讨论两种“联合作战范式”以及三次“军事革命”中的头两次时，拿出来做进一步的阐述。海战、空战和太空战出现较晚。到了第二次世界大战，这四个战争领域已经合到了一起。最后，关于军事变革的性质，书中将从三个角度进行阐述：研发、军民两用和军事革命。

本书中所使用的“作战”（warfare）一词，指的是针对敌人所采取的战争行为。它使用或威胁使用武力去杀死、俘虏或强迫敌人，以实现己方的意愿。因此，作战一般是在战争状态下采取的行动。按照马克斯·韦伯的经典定义，战争（war）是国家之间有组织的武装冲突，而国家则是指在自己的领土上垄断了武装力量的政治实体。后来由于众多的非国家性质的角色卷入类似战争的行动，人们习惯于把战争定义为存在于社群之间的一种状态。就我们讨论的历史而言，韦伯的定义是适用的。战争是状态，作战是行动。

“技术”的含义则没有这么清楚。本书谈及的技术是指人类对物质世界的一种有目的的操控。它通过技巧性地使用工具或机器，运用动力来导致某种物质的改变。就其本质而言，技术是为达到某种人类目的而改造物质世界的一个过程。对想法、观念、感情、关系、信仰、情绪等人类精神层面的操控，或许是一些技术的次级效果。但如果物质层面没有发生改变，那还不能算是技术。总之，技术是人类活动中最为物质化的。作战也是如此。实际上，作战和技术两者都能塑造甚至决定战争的结果。但它们都不是战争本身。按照克劳塞维茨的说法，战争不过是用其他的方式延续的政治。作战（包括其中用到的技术）也是如此——它们也不过是用其他方式延续的战争，而且所用的方式具有广泛深远的物质性。

有些打着“作战”旗号的行动其实未必名实相符，比如我们后面将讨论的“网络战”。虽然也使用技术改变了物质世界，但还不能提升到作战的级别。恐怖主义并非战争的一种形式，它只是战争中可能用到的一种手段，或者是展示个人怒气或愚蠢的一个工具。人们会对恐怖分子宣战，但不是对恐怖宣战。心理战所操控的，更多是人们的想法而非物质。这里会用到技术，但不是根本性的。

最后还有一则定义需要引起重视。人们一般会把技术产品视为技术本身，像是航空母舰、坦克和轰炸机之类。这些技术产品是实现航行、射击和轰炸能力的技术体系的一部分，但它们本

5

身不是技术。就本书而言，这个区分很重要。因为在作战历史上，防御工事和道路，都位列最为重要的技术产品之中。这类技术产品以及产生它们的技术，会在本书后面部分经常出现。

本书所研究的主要是西方的历史，也是作者最为熟悉、史料最为丰富的那一部分。但本书提出的观点和概念不限于此，是普遍适用的——这也是本书存在的前提。

6

第二章

陆 战

史前作战

对于人类文明曙光出现之前的技术和作战，我们证据不足、所知甚少。不过，从史前的迷雾中，我们可以依稀辨识出一些模式。20世纪90年代，人们在德国黑尔姆施泰特露天开采的褐煤矿发现了一条惊人的线索。依据该矿命名的“舍宁根工程”出土了多达11件木制标枪——它们在原先是个湖泊的沉积层中保存了三十万年。杉木和松木，被制成不规则的、带尖端的木杆，长度为5.9—8.2英尺。最为引人注目的是，它们的杆体由粗到细，跟现代标枪一样前重后轻，利于飞行。如果由海德堡人投掷这些标枪的话，可以达到35米远。

这些人造器具，告诉了我们许多有关史前武器技术的事情。首先，早在智人出现之前的十万到二十万年里，原始人就开始利用自然条件了。大量人造器具的证据表明，人类曾使用石头和骨头作为武器。在中石器时代和新石器时代，石头被磨成便于

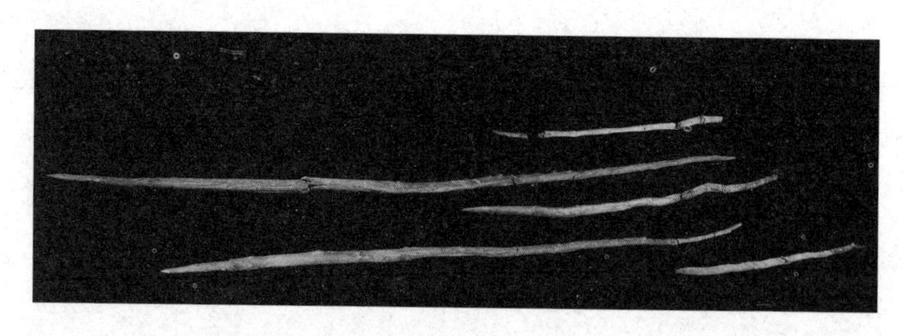

图1　迄今为止发现的最早的武器制品。这些舍宁根标枪是军民两用的手工制品，可用于狩猎，也可用于作战。它们的使用者，是三十万年前生活在欧洲中部的海德堡人。它们证明了人类是带着武装来到世间的

使用的形状。我们也可以合理地认为，他们应该也会用类似工艺处理木头——尽管这些做好的木制器具后来大多朽烂了。这些舍宁根标枪无可辩驳地证明：更早的时候，就有了复杂得多的木制工艺。我们可以根据标枪证据推测，在跨越石器时代的数千年里，就已经有了木制的矛、扎枪、棍棒，甚至刀。我们不能确定，这些标枪和其他武器是用来打猎还是作战——或许两者都有；但根据这些，我们可以更有把握地认为，智人是带着武装来到世间的。

考古发现未能回答的另一个大问题是：石器时代的武器是用来狩猎的，还是用来作战的？抑或两者都有？现存的可靠证据，如骸骨、石头和岩画，大多来自中石器时代后期或新石器时代，距今大约两万至六千年之间。在此之前，当时的武器显然已经同时用于狩猎和作战。没有理由认为：一种武器仅限于其中一种用途，而不做他用。考古发现的那些涂了毒素的箭头，或许仅限于攻击那些并未打算食用的目标。但大多数史前武器——投石器、矛、扎枪、棍棒、刀、斧、狼牙棒、梭镖投射器、澳洲标枪——都是我们今天称之为“军民两用”技术的早期案例。这

些技术既可以用于军事用途，也可以用于民用目的。可以想见的是，这里面的一些武器是为作战而发明的，只不过后来又转到狩猎用途，反过来也是一样的情况。 8

这种概括同样适用于史前最伟大的军事技术——弓箭。弓箭发明于四十多万年前的旧石器时代，之后在狩猎和作战中不断使用，一直到今天。其他的史前武器都是工具，弓箭却是一种机械。它拥有可移动的部件，并能够存蓄能量。其他的史前武器很直观，而弓箭的发明需要想象力的飞跃——需要有想象出一件自然界并不存在的东西的能力。我们无法知道，这项堪称奇迹的发明，是突然发生，然后渗透并传播到世界各地的人类群体的，还是由本地的天才反复改进才得以出现。古希腊和古罗马人当年想象出的掌管武器制作的神，都是一位打造金属器件的锻造者——古希腊是赫菲斯托斯，古罗马是伏尔甘。其实真正的武器制作之神，应当是那位在旧石器时代发明了弓箭的“爱迪生”。

尽管人们还未能解开史前时期武器的很多秘密，但仅凭掌握的有限知识，我们可以对技术和作战的根源做出一些概括。首先，如前文提到的，这些致命的技术大多是军民两用的。其次，它们既包括投射武器，也包括击打武器，这种差别一直延续到了今天。使用投射武器是在一定的距离之外，这有助于猎手或战士免受伤害。而那些用来打或刺的击打武器，虽然更为致命，却需要使用者与目标进行接触。如果击打目标碰巧是史前人类所喜欢的大型猎物，或者是另一位战士，这场遭遇战就会变得很危险。这个难题一直贯穿了整个人类历史——从号称“黑人拿破仑”的南非祖鲁族的恰卡（他把传统的投射标枪改造成 9

了短矛)，到子弹打光后不得不插上刺刀进行肉搏的现代士兵。

投射武器和击打武器的对比，也引出了史前狩猎和作战的第三个特点。通过研究19和20世纪那些使用史前武器争斗的社会，我们能推断出："打了就跑"的战术通常是最好的选择。因为大型动物和对方的战士都很危险，所以要杀死目标，最好的方式就是伏击。通过投射或击打行为，出其不意地攻击并重创目标，然后撒腿就跑。如果攻击成功，晚些时候就可以回来收拾尸体或处决伤者。从历史上看，当相对较弱一方面对强大的敌人时，伏击是这种非对称作战的首选技巧。到了21世纪，简易爆炸装置成了新时代的"舍宁根标枪"而被用于伏击。

古代作战

公元前十千纪至公元前四千纪中期，新石器时代的革命来到地中海东部的黎凡特地区，并在此不断扩展。在这六千年里，这一地区的居民学会了驯化植物和动物，在河谷定居了下来。他们建造村寨，村寨进而发展成城镇。一些驯养动物的早期村民，搬离河谷，住到了周边的高地上，并利用农耕群体和继续漫游的采集狩猎群体之间的中间地带放牧畜群。所有这三类人类群体——觅食者、畜牧者和农耕者，都由于内部或社群间的争斗发展出了作战技术。

在文明早期这些定居的农耕者中，产生了古代最为重要的军事技术——筑造城防。其他的军事技术，有助于决定战斗或战争的输赢；而城防工事，则有助于决定一场战争或战斗到底会不会发生。农耕者驯养植物和动物，形成了定居社群，在满足基本需求之外，开始有了积余。简单的住所里，有了余粮、衣服、首饰、炊具、餐具和家具。房子也变大了。而游荡乡间的掠食者——动物或人——会袭击这些食物的集中地，将它掳掠一空。插在地里的一根根挨着绑起来的木桩，构成了最简单的防护篱笆。随着这些简陋房屋和围墙里的木头被土和石头取代，新的筑造技术出现了。这些技术后来发展成为宏伟的建筑设计，为城邦打下了基础。一个很好的问题是：是先有了筑造城墙的技术，然后再被运用于住家和公共建筑，还是反过来的演化过程，即先造祭坛和神庙，再用同样的材料和工艺加固他们的围墙？不管怎样，能够建造宏伟长久的公共建筑的这种军民两用技术，成为早期伟大文明的象征。实际上，我们所说的"文明"这个词，就来自罗马语里的"城市"一词。

杰里科古城，作为最早的样本堪称久远——无论就其地理位置还是时间顺序而言，都是如此。同新石器革命时期其他的定居点一样，杰里科很早就开始驯化植物和动物。但不同的一点是，它加固了自己所处的位于死海以北、约旦河谷的这块绿洲的城防。到了公元前8000年，一个占地约4万平方米、拥有2000—3000人口的城镇，就处在5英尺厚、12—15英尺高的石头城墙保护之下。顺着一处城墙，有一座距地面28英尺高的塔，通过里面的楼梯登上塔顶，放哨者可以瞭望周围数英里田野的情况。石器时代的杰里科居民没有留下文字记录，来说明他们是谁，以及他们建造这个前所未见的防御工事的方法和原因。

这个镇子坐落在几条贸易通道的交会处，这意味着它可能会受到来往人员的劫掠。我们不知道确切的情况，但考古研究表明：这些城墙直到公元前16或15世纪才开始颓圮——这发生在《圣经》记载的时代之前。总之，杰里科古城的城墙耸立了

图2　耸立在死海北面平原之上的杰里科古城（今又称“苏丹废丘”）是最早的史前城防样本，也是塑造了人类历史上战争和作战方式的非武器技术产品。这张照片显示了杰里科古城墙的局部，以及12英尺高塔的顶部——在那里曾经可以瞭望到周边平原

12

六千多年。在此期间，虽然城里的居民换了两拨，但始终没有被外来的暴力所征服。

与杰里科相比，位于幼发拉底河畔的乌鲁克古城，让我们可以更真切地了解人类早期的宏伟要塞。乌鲁克兴盛于大约公元前2900年，属于青铜器时代中期。现代考古学对这座宏伟的城邦已经有了可靠的发现，我们可以把它们与文字记载的城市创立神话加以对照。来自这两方面的信息，揭示了宏伟建筑所运用的技术及其在社会中发挥的作用。

同《荷马史诗》一样，《吉尔伽美什史诗》最初是口口相传，后来才被那些未曾亲身经历事件的作者用各种文字形式记录下来。关于史诗的源头，学者们通过数十载的研究形成了广泛的共识，并写出浩繁卷帙阐述其含义。我们基本可以确定：吉尔伽美什是一位真实存在的人物，他在公元前三千纪的前半期曾统治过美索不达米亚的乌鲁克王国。较真的读者需要从他的传奇故事里去伪存真，不过即便是那些传奇，也是有启发性的。三分之二为神、三分之一为人的吉尔伽美什，统治了乌鲁克一百二十六年。这部史诗，讲述了他为追求名声和永生而进行的英勇征战。其中的一次征伐，是来到了芬巴巴统治的雪松森林。芬巴巴是一位堕落下界的神，掌管着通往地狱的地下河。这个雪松森林，很可能就在今天土耳其的努尔山区，位于幼发拉底河的上游。与吉尔伽美什一起征战的恩奇都，是一个被文明所诱惑的野人。当他们遇见芬巴巴时，吉尔伽美什用神为他打造的兵器杀死了芬巴巴，但芬巴巴临死前的注视，让恩奇都成了受害者。为寻求永生，吉尔伽美什曾往返地狱。他归来时终于明白，自己终将步恩奇都死亡的后尘。

虽然《吉尔伽美什史诗》探讨的是爱情、生命和死亡等精神主题，但它也让我们更多地了解了青铜器时代美索不达米亚地区的物质世界。吉尔伽美什去雪松森林是为获取木头，用来建造乌鲁克的城门，也许还用来烧制建造城池的黏土砖。吉尔伽美什经常吹嘘，自己的城砖烧制得坚硬，只有那些足够富有和勇敢的人才能奢侈地拥有那么多的木头。在吉尔伽美什统治时期，他的城墙号称有3.4英里长、占地2.3平方英里——面积大约是杰里科古城的220倍。吉尔伽美什时代的人口超过8万，是当时世界上最大的城市。环绕城墙的一条护城壕，为入侵者设置了另一道障碍，同时还能溺毙那些企图从城墙下面挖地道进来的人。在这些25英尺厚的城墙里面，分布着高大宏伟的民用建筑、神庙和其他公共设施。它们提供了社群活动空间，也让目睹者心生敬畏。在称呼他的所有头衔里面，吉尔伽美什最为看重的是“城墙建造者”这一项。他所精通的这门技术，保障了他的城市的安全和繁荣。较之于他神勇的兵器，吉尔伽美什的名声更倚重他所建造的牢固要塞——“坚不可摧的乌鲁克”。当然，到了这个时期，史前的作战，显然已经变成了马克斯·韦伯所称的国家之间有组织的武装冲突。

在高大的城防工事改变诸如巴比伦和尼尼微这样的城市冲突形式的时候，另外两项军事技术也在改变野战方式。正如“青铜器时代”名称之所指，金属在这个时期很快取代了石头，14 出现在了箭头、矛、刀等用来刺穿人和动物肉体的器具上。不仅如此，青铜的应用也导致了一种全新武器的出现——这就是剑。早前用来刺击的武器，是用石头和骨头制作的，其重量和脆度限制了武器的长度。青铜，是铜和锡的混合物，可以做出数英尺长、两面锋利的刃，便于刺击或砍击。在最早的文字记载中，上古文明的神话英雄具有神力，能够挥舞神仙赋予他们的武器。比如，吉尔伽美什就有一张用名贵的安善木材制成的弯弓，以及众神为他打造的号称“英雄之威”的斧头。据说，他的武器有600磅重，普通人根本拿不动，只有半人半神的吉尔伽美什才能运用自如。较之于世上其他的军事技术，剑，通过传奇故事、民间演义和神话传说，迅速拥有了象征意义。从亚瑟王的神剑到日本的本庄正宗（日本名剑），剑都为作战平添了浪漫的味道。世界各地的军人，在阅兵式上依然保留佩剑，就是对过往时代的记忆。那时候的战士们相信——或者愿意相信——某些兵器能够将美德、正义、荣誉甚至虔诚，转化成战斗的胜利。又或者反过来，也许神奇的剑本身，就体现了神的恩宠。现代的战士们，依然给他们的剑取上各种名字，比如天神宙斯、爱国者、圣骑士、和平缔造者等。

青铜器时代最伟大的武器发明，并非来自这些新的文明，也不是由青铜打造。公元前17世纪，在欧亚大草原上发展出来的木制二轮战车，不期而至地闯入了黎凡特地区。早在公元前四千纪，美索不达米亚地区就有了战车，但这些由驴和牛拉动的战车很笨重，用的是四个固定的实体车轮，只能以缓慢的速度拉着士兵和装备到达战场。相比之下，二轮战车则可以在战场纵横驰骋。它由两匹或四匹马拉动，只用两个辐射状车轮，能够迅速地追击敌人。它横扫战场，通过穿插或包围步兵队伍，迫使敌人要么崩溃投降，要么用同样的战车武装自己。在差不多六百年的时间里，历史学家威廉·麦克尼尔所称的这种“古代超级武器”，使得当时的强国和有意争霸的国家不得不加入一场前所未有的军备竞赛。为了竞争，各国即便自身条件不足，也不得不发展木工、汇集马匹，为国防和对外征伐建造军械库、马厩和维修所。对这些新奇战争武器的需求如此巨大，以至于出现了一个国际雇佣兵阶层——战车武士，他们专门向那些没有掌握这项技术或无力维持常备战车的国家提供士兵和装备以获取利益。据说，所罗门王拥有的战车达1400辆之多。

根据一些记载，规模最大的战车大战发生在公元前13世纪初，在今天叙利亚奥龙特斯河畔的卡叠什古城之外。对阵双方分别是埃及国王拉美西斯二世和赫梯国王穆瓦塔利斯。围绕这场大战说法不一，但大家公认的是：双方动用了数千辆战车和数万名士兵，战斗正酣时，拉美西斯二世陷入险境，结果不得不撤回埃及。这场时代决战的天平倒向了半开化的赫梯人。

尽管二轮战车堪称史上最重要的武器发明之一，我们却并不确切地知道它在战斗中如何使用。大多数人把它视为一个作战平台：战车上有一名驭手，负责驾车冲向敌阵，车上的另外一至二名乘员，负责使用投射武器——向敌人射箭或投矛。还有一种用法是把战车当作运输工具，将战士中的精英运送至战斗地点，然后下车徒步进行战斗。这种用途在青铜器时代末期的《伊利亚特》里有所描述。比如，阿喀琉斯就是被运送至特洛伊城下，在那里与赫克托进行肉搏。阿喀琉斯在杀死这位特洛伊的英雄之后，用战车拖着赫克托的尸体环绕城墙。战车的第三个用途，是发挥冲击作用——战车直接冲入敌军步兵阵型，车上的弓箭手和投矛手在穿插中向敌人射击。

不管它是如何使用的，这种战车在黎凡特地区出现没有多久就很快衰落了。大约公元前1200年之后，二轮战车失去了在

图3 公元前二千纪出现的二轮战车，是作战方式的一场革命。图片刻画的是公元前1274年的卡叠什战役，埃及法老拉美西斯二世驾驶战车碾过倒在他弓箭之下的赫梯士兵。这种运用了军民两用技术的战车，是史上最早的陆地作战平台

黎凡特地区作战中的主流地位，并且一去不返。随后几个世纪里，这种技术向东和向西迁移，在印度、中国、希腊、罗马、欧洲大陆，甚至英国和爱尔兰都曾出现过。但最终，它还是从这些地方的战场上消失了，退归到打猎、典礼、交通、体育（比如赛车）等用途。那么，是什么原因导致了如此威猛的作战系统迅速陷入衰落？因为公元前1200年大约是青铜器时代让位于铁器时代的时间，据此一些学者猜测，是新出现的铁制武器让步兵有了抗衡二轮战车的能力。但这种解释并没有得到多少人的认同。除此之外，还有一种从经济角度进行的解释：围绕二轮战车的军备竞赛代价太过昂贵，使得所有参与者最终都耗尽了财

17

力。其他一些学者则认为，变化的一个原因在于被称为“大灾变”（Catastrophe）的事件。公元前1200年前后，可能由于环境和气候改变，来自欧亚大草原的一些野蛮部落迁移到靠近黑海、爱琴海和地中海东部的西亚地区。他们在迁移过程中赶走了这片土地上的原住民；而后者转而攻击南边的邻居，从而造成了连续的被迫迁徙。一波波入侵的浪潮，在公元前13世纪，由“海上民族”登陆埃及达到高潮。当时，拉美西斯三世乘着他的二轮战车迎战这些水陆两栖的入侵者——这大概是二轮战车这种超级武器在黎凡特地区最后一次华丽的亮相了。

二轮战车究竟为何衰落？一个似乎符合所有证据的解释是：由于步兵采用了一种新的战术。这种战术或许是发动“大灾变”的草原勇士们所带来的。因为他们自身就是骑士，深知马是不会往一堵墙上撞的——那么面对牢固的人墙也应如此。如果二轮战车是用来冲击队列的，一旦当时的步兵意识到，只要凭借手里新出现的铁制武器坚守阵型就能阻止对手，那么二轮战车的威力瞬间就能被瓦解。作为一种震慑武器，也许它可以一直存在，但其影响更多局限在心理层面而非实际威力。

不管怎样，二轮战车在黎凡特地区的作战中不再是主角，开始退到后方发挥辅助性的作用——用于运输、典礼、马戏、打猎之类的场合。二轮战车主宰西方作战方式的时期，虽然短暂却引人注目，并由此产生了一次军事革命——这也是本书重点讨论的三次军事革命中的第一次。这里的“军事革命”，是指那些对作战方式影响深远的变革。它不仅重新定义作战性质，而且通过改变国家之间的关系和获得强权的手段，从而影响到历史进程。正如威廉·麦克尼尔对二轮战车的评价，它“彻底改变了欧亚地区的社会平衡”。二轮战车，不光影响巨大，还带来了在技术和作战发展史上反复出现的若干问题。

第一，这确实是一件革命性的武器。在它所到之处，所有国家都不得不接受它、反制它，或者与它的拥有者达成和平。按本书的术语，战斗选项可分为对称性的和非对称性的作战。非对称性选项的采用者，必须拥有反制的技术或技巧。但是在六百年的时间里，似乎还没有人有能力阻止它。相反，人们接受了它——这也是衡量一项真正革命的尺度之一。也就是说，他们选择了对称性作战，就像20世纪冷战时期的军备竞赛那样。第二，二轮战车是由野蛮人而不是文明人发明的。历史上军事技术的发明，多以文明为中心，因此文明人在对阵野蛮人的时候通常也占据优势。但在此案例中，却是来自欧亚大草原的野蛮人引发了这场技术革命：他们先是驯化了马，然后又给它们套上了用于作战的车辆。而那些文明，一旦与这种新技术遭遇，只能纷纷拥抱它，否则的话，只有臣服。

还有第三点。直到公元前1200年二轮战车式微之际，人们似乎并没有找到可以抗衡它的技术。也许有的军队会利用地形使得二轮战车无法发挥作用，但当时并没有出现反制它的武器，作战依然是对称性的战车对阵战车。第四，二轮战车技术传播很快，从欧亚大草原传到黎凡特地区，随后又传给了欧亚的大部分文明国家。军事领袖在面对它时，必须迅速做出抉择：是拥有它，还是臣服于它？第五，一项技术发展受阻，通常要么源自自身的内在缺陷，要么是被能反制它的新发明所打败。所以，二轮战车走向衰落，原因要么是在于它的高成本，要么是被对手掌握了打败它的技巧。

19

第六，同剑一样，二轮战车也蒙上了跨越文化的象征色彩。古埃及的法老们喜欢让人绘制自己驾驶战车打猎或作战的形象。无论平民或军事领袖，要展示威严、权力和胜利的气场，都会选择乘坐二轮战车。第七，二轮战车开启了骑兵和步兵的交替轮回，这种轮回一直延续到了21世纪。特别是在西方，有史以来的重大战事中，作战主导方式不是骑兵就是步兵，每种方式所运用的武器系统都主宰了当时的战场。从大草原呼啸而来的二轮战车，开始了第一轮的马上作战，而它的衰落又使得黎凡特地区回归到步兵作战的方式。本书的后面章节将探究并试图解释，推动一个轮回走向下一个轮回的各种力量——尤其是技术。第八，按照同样的发展路径，二轮战车不是简单的军民两用技术，它兼具五种用途：作战、运输、打猎、典礼和体育。欧亚大草原上的游牧民族最初可能只是用它来打猎，但其他人发现了它更多的用途。第九，也是最后一点，二轮战车是第一个用于地面作战的武器平台。直到20世纪坦克出现以前，在陆战中它一直是无与伦比的。就组织和军事功能而言，它堪称军舰、军机和航天飞机的先驱。同其他领域的平台一样，二轮战车要求它的一部分乘员操控车辆，另一部分乘员操作武器。这一点，遥遥领先于它那个时代。

第一联合作战范式

公元前1200年前后的“大灾变”，让黎凡特地区进入了一个“黑暗时期”，经济、政治、军事和技术陷入停滞。二轮战车衰落之后，陆地作战除了联合作战范式外并无创新。直到中世纪晚期，那些文明国度里的军队作战，都是依靠重装步兵结成方阵，再由骑兵和轻步兵提供支援。各国武装力量的构成，在差不多一千年的时间里，大体都是如此。重装步兵手持矛或剑。矛，可以是罗马标枪那种投射武器，也可以是马其顿那种超过20英尺的长矛。剑，可以是罗马短剑，也可以是萨珊长剑。在近战中，士兵也会使用其他各种刺击或打击的武器。这些重装步兵，披盔戴甲，手持盾牌，身体部位各有防护。他们大多有头盔、护胸甲，有的还有护胫板（肉搏中保护胫骨不被踢伤）等其他特殊防护设备。

轻步兵为重装步兵提供支援的方式，是从侧翼或正面（如在遭遇战中）投射武器。他们使用弓箭、标枪或投石器。因为他们依赖机动性来获得保护，所以他们的防护设备即便有也很少。骑兵使用战车、马或骆驼。这一时期记载的为数不多的卷镰战车，显然是被用来冲击敌方战阵的。不过，战车也许还被用作投射武器的发射平台，或者用来发挥阻隔、侦察等功能。骑兵发挥的也是同样的作用。

在这种联合作战范式中，步兵一直占据主导地位，一直到罗马帝国的没落。此后轮回交替，骑兵主导的时代再度来临，并延续到火器革命的时代。无论是亚述人、波斯人、古希腊人、马其顿人、古罗马人、萨珊人，还是来自森林和草原的野蛮人或来自沙漠地区的穆斯林，虽然作战方式各有不同，但其差异也仅仅是在组织、战术、战略和文化等方面。在步兵和骑兵的轮回交替过程中，古典时期到中世纪时期的陆战技术，本质上都差不多，锁定在了一种停滞的野战范式。其间，各个国家依据其自身的财力、自然资源、劳动力和兵员来配备武器装备，并选择适合自己的战争方式。

21

新亚述帝国

一个与上述模式显然不同的例外现象，格外引人注目。新亚述帝国（公元前911—前612）是个全面军事化的国家，也是史上记录的第一个掠夺国家。在长达三个世纪的时间里，新亚述人四处扩张，贪婪冷酷地不断攻伐别国。他们尤为见长的是野战方面的刻意创新——特别是攻城战。尽管是内陆国，他们也建造了战舰。根据史料描绘，他们全副武装，利用充气的动物膀胱涉水渡河。他们还修筑道路，来连接不断扩张的帝国版图。他们的士兵装备着最新式的、高质量的军装、盔甲和武器。他们还复兴了战车，将它改造成车轮为十二轮辐的重型车辆。新的车型可以载四名乘员，甚至能深入美索不达米亚河谷的丘陵和山地等恶劣地形。这种更重的战车也能够直接冲向敌人的步兵战阵，实施冲击战术。有证据显示，亚述人最先引入卷镰战车——这种战车专为砍倒阵列中的步兵而设计。所有这些围绕战车的创新，都是与新亚述人侵略性、血腥和恐怖的战争方式相一致的。

新亚述人在攻城战上的创新则更为显著。除了传统的攻城工具云梯之外，他们增加了带轮子的攻城塔，这样己方士兵可以直接近距离攻击城头的敌方防守者。他们还发明了攻城锤，一种用来撞击城门，另一种用来撞击城墙。用于城墙的攻城锤也有两种版本，一种用来重击城墙，另一种专门对付美索不达米亚地区修筑要塞常见的黏土城砖。为了破城，他们还在城墙下面挖地道，并且在城外架设投石机。

然而，新亚述时期强盛的军事技术——无论是野战还是攻城战——并未能终结自“大灾变”以来西方作战技术所陷入的停滞状态。即便是他们独创的攻城装备，也改变不了复杂坚固的要塞所具备的优势。他们可以包围城市、饿死城内居民，也可以给城内水源投毒，还可以屠杀陷落城市的居民来恐吓其他试图抵抗的城市。他们甚至可以利用诡计来攻破城池，就像古希腊人使用的特洛伊木马那样。尽管萨尔贡二世（约公元前721—前705）和其他亚述国王都喜欢标榜自己“城池摧毁者”的形象——这与吉尔伽美什引以为豪的“城池建造者”名号正好相反——其实并没有多少证据证明，他们的攻城技术征服了很多城市。马其顿以围城著称的国王德米特里乌斯，在公元前305至前304年曾花了一年的时间围攻罗兹城，却没能攻破。他在战场上使用了当时最大的攻城机械：高达九层的可移动攻城塔，塔上有好几层还装有投石机。但如此了不起的工程装备，还是输给了罗兹城众多的投石机。正如历史学家保罗·本特利·科恩对此次战役的评论：“古代的攻城作战，已经走到了技术的死胡同。要摆脱困境，需要等到一千五百年后火药的出现。”围困是亚述人及其后来者征服城市的传统做法。这种无论野蛮部落、弱小城邦，还是强大帝国都沿用的做法，经久不衰，一直到了中世纪。1453年，随着君士坦丁堡城池陷落，攻城技术才有了突破。

但是，新亚述帝国在技术上的独创性和有效性却不容否认。用什么可以解释这种耀眼的技术创新？仅仅因为新亚述人比同时代的人对知识更为好奇吗？他们的国王亚述巴尼帕（公元前668—前627）建造了无与伦比的图书馆，似乎印证了这一假设。还是说，因为他们人口数量小于他们的野心，所以追求省力的机械，而军力借此也得以提升？或者是因为，所有穷兵黩武的国家都对新武器和新装备孜孜以求？不管是出于何种缘由，新亚述人带来了大量新的军事技术。

不过，这些创新并不能保证成功。更确切地说，它们倒是引发了攻城战中“相互克制”的技术模式，并且这种模式还一直延续到了现代。守城者建造坚固的城墙，攻城者发明攻城塔来攀爬城墙；前者往攻城塔上放火，后者用蘸水的兽皮来防火；一方用抛射武器来击破城墙，另一方则在城墙上安放类似武器来攻击攻城者的机械——如此这般你来我往。不同于仅仅用于反制的技术，技术上的相互克制，引发的是机器般不断相互激发的创新模式。在第一联合作战范式时代，守城往往比攻城更为成功，这与新亚述帝国所吹嘘的恰恰相反。

古典作战

公元前612年，亚述帝国衰落，随后西方文明很快结束了我所称的“古代时期”（公元前3500—前500），进入了古典时期（约公元前500—公元500），也就是古希腊和古罗马时期。古希腊人和古罗马人依然采用第一联合作战范式，但改进了亚述人的攻城机械等军事技术，他们在文字记录、官员体系、道路交通、城防要塞等方面也有改善。在此过程中，他们显然以现代形式开创了工程学。亚述人很可能有自己的工程兵，他们留下来的图画和文物依稀有所反映。但是，只有古希腊人和古罗马人留下的文字记录，才证实了古希腊和古罗马的工程所达到的高超水平。

24

西方文明研究中对此论述甚多，本书则从公元前一千纪中叶的古希腊讲起。与同时代人的文明追求相比，古希腊城邦的居民们更倾向于用理性的方式阐释自然界。作为城邦的装饰，他们还培育了哲学、科学、政治、文化和艺术。一些学习西方文明的学生发现，西方文明根植于他们所谓的“古希腊奇迹”。在军事领域，一位历史学家甚至声称，是古希腊发明了“西方的战争方式”。尽管大多数学者对此见解不敢苟同，但得到广泛认同的是，古希腊文明给世界带来了众多的理念、信仰，以及思考和感知的模式，这些组成了西方的世界观。

古希腊在军事技术领域最重要的贡献，按今天的说法，是以科学为基础的工程。也就是，依靠数学（现代语言谓之“科学”）进行机械和结构的设计、建造和使用。古希腊人被证明特别擅长攻城技术，以及它的反面——孪生相伴的守城技术。古希腊的思想和器械在地中海地区广为传播，特别是在后来的罗马共和国和罗马帝国扎下根来。在那里，军事技术显得尤其重要，其光彩在很多方面都超越了罗马军队盛名之下的野战能力。

古希腊人和古罗马人借此给世人留下了一整套攻城机械。除了原有的攻城锤和移动攻城塔，他们发展出了不同形式的投石机：石弩、弩车、弩炮、蝎子炮等。这些后出现的投射器是现代大炮的前身，所储存和释放的能量都来自拉力、扭力和重力这三者之一。拉力和扭力机械，分别是拉伸和扭曲有机物质，比如绳索、木头、动物毛发或筋腱。投射机械，按照抛物线投出物体，可用来向敌人的阵地发射火球或动物尸体、蛇等让人不适的东西，但要对城墙造成很大破坏却并不容易。直接火攻，也许能造成城墙破损，甚至能在城门附近或城墙薄弱处打开缺口，但其威力并不足以撕开城防——即使面对美索不达米亚地区的土砖城墙也难以做到。用攻城塔攀越城墙，或者在城墙底下挖地道，虽更有希望破城，但其有效性也受到火油防守和护城壕的限制。所有这些精巧的装置，很可能更多地只是发挥其在心理层面的作用。攻城战中最有效的形式依然是非技术的手段，如谈判、饥饿、恐怖、诈术和策反。

除了精巧的攻城机械，古典时期的军事工程师们还做了很多其他的贡献。首先，他们给君主们灌输了技术能带来军事优势的理念。有些工程师在宫廷任职，还有些活跃在地中海地区，他们四处兜售自己的技能。叙拉古的狄奥尼修斯一世甚至专门设立了一个军事研发中心，据说是用来研制投石机。阿基米德是当时最杰出的数学家，死于罗马对叙拉古的入侵。被杀死之前，他还在研究如何用镜子聚焦太阳光来烧毁侵略者的舰队。他也许还发明了一种使用杠杆原理的吊车来倾覆敌舰，但我们无法确认这类故事的可信度。就他和古希腊时代同事们的发明而言，虽然非常聪明，但实用性有限。

工程师们对我们今天称之为“民用工程”的贡献，更令人印象深刻，并且可以验证。罗马围绕地中海铺设了5.5万英里的主次干道，借此可以在帝国范围内迅速调遣罗马军团。这些道路施工精妙，标准的计划同当地的材料和地形相结合，因地制宜地铺就了宽深不一、坡度平缓的笔直大道。道路连接处，有木桥或石桥衔接。作为军民两用的技术产品，罗马大道不仅为国家的军事和战略目标服务，而且有助于国家治理、商旅往来，有效提升了国家的凝聚力。古代的波斯、亚述和中国，包括后几个世纪出现的印加帝国等国家，也都建有类似的商业、军事和政府通信用途的国家干道，但都无法与罗马的道路系统相提并论。直到德国的高速公路和美国的州际公路建成，它才算被超越。

罗马军队的士兵建造了很多干道，也把同样的技能和知识运用在战役之中。渡河时，他们采用浮桥。恺撒大帝面对日耳曼部族，曾两度在莱茵河上架设木桥，最终让对手明白：抵抗是徒劳的。罗马军队喜欢采用的攻城技术，也体现了同样的思路——他们把土一直堆到敌人城头。这种古老的攻城术，在罗马人这里得到发扬光大。他们在马萨达古城留下的土堆还在那里，附近负责施工保护的古营地遗址，到今天依然清晰可见。罗马共和国和帝国军队的战事失利，多半是由于指挥不当。比如，汉尼拔几次大败罗马人，都是发生在罗马本土。但罗马人总会杀回来，坚持不懈直到取胜为止。军事工程，对他们来说，不仅仅是展示国家力量的一种工具，它成了一种国家理念，一种战争方式。罗马人抛弃了希腊人那种文雅的、建立在数学基础上的军事工程，他们拥抱的是一种实用主义的、不断尝试的工程技术，其中很多是在战场上学会并加以总结后传承。他们的不少敌人最后屈服于罗马，不是被打败，而是实在疲于应战。

尽管罗马在工程上建树颇多，但古典时期的野战并未摆脱第一联合作战范式的窠臼。最显著的创新，也仅仅局限于一些静态技术形式的变化。罗马步兵使用的西班牙短剑，就是一个例证。短剑，是罗马军团的标准配置，它比一般的剑都要短，但在长度、刃型、剑柄，特别是在材质上，随着时间在不断演变。这种变化表明，他们非常善于吸收新的军事技术。正如古希腊历史学家波利比乌斯所言："没有人比罗马人更愿意从别的民族那里学习新的时尚，模仿别人优于他们的长处。"在第二次迦太基战争期间，罗马人就发现，伊比利亚人使用的剑，比他们自己正在使用的铸铁剑，有不少独特之处。汉尼拔·巴卡，从他在伊比利亚半岛迦太基的家族领地发兵攻打罗马。交战中，罗马人很快就发现，来自西班牙的剑比他们的硬，而且剑尖和侧刃更为耐用。这些剑可以磨得如同剃刀般锋利，在战斗中大显神威。罗马人仔细研究了伊比利亚的制剑工艺，并将其带回了老家。但是，他们没有办法把伊比利亚的铁矿带回去，而其实正是那里的矿石造就了托莱多钢的特性。因此，罗马人的仿制品很少能做出真正西班牙短剑的品质。后来，罗马人又以剑的出产地——美因茨和庞贝——来命名他们的短剑，但都不如前者的名头响亮。正如先前传说中的天神赫菲斯托斯和伏尔甘为吉尔伽美什打造斧头这类神奇兵刃一样，西班牙短剑也成为罗马传奇的一部分。姑且不论这类武器是否具有超自然的威力，单就它们存在本身就赋予了使用者“天命在身”的光环。

复合反曲弓成为轻骑兵的首选武器，是第一联合作战范式在古典时期发生的第二项重大变化。同之前的二轮战车一样，这种军事装备也是野蛮人发明的，可能早前在欧亚大草原上就出现了。这是一种短弓，弓的两端向外反张，特别适合在马背或战车上发射——骑手或乘员可以很容易地避开马头或战车围栏，左右开弓。与西班牙短剑一样，它也需要特别的材料和制作工艺，来保证它的独特性能。它的层压结构，令它总体上可以发挥最大的强度和威力：通常上面一层是动物筋腱，中间是木头，后面一层是动物的角。各层用胶水粘连之后，加以蒸烤弯曲到特定角度，然后再包裹定型。这种弓，不上弓弦就不算长，上了之后会更短，非常便于携带，射击时也便于操作。在娴熟的弓箭手手里，它可以发挥巨大的威力和很高的精准度。

因此可以说，在第一联合作战范式内部出现了技术变革，但作战范式本身依然停滞不前。在欧亚战场上，基本的武器装备花样百出，再结合各自的军队，可谓各领风骚。但纵观古希腊、马其顿帝国和罗马共和国的起起落落，西方军队的中心位置始终为重装步兵所占据。一代又一代结成方阵的披甲步兵，是军队的王牌力量和陆战搏杀中的威猛之师。而古希腊和罗马在西亚遭遇到的敌人，其军队多为轻骑兵或重装骑兵——或者两者都有。西流基人、帕提亚人、亚美尼亚人、斯基泰人、萨珊人等西亚王朝所采取的野战方式，要么是由集结在一起的轻装骑兵使用反曲弓射击，要么是重装骑兵（马有时候也披甲）持矛冲击。 29

重装骑兵一般由贵族组成，一个社会中也只有他们能负担得起作战需要的全套甲胄。在步兵的全盛时期，纪律森严的重装步兵，在对抗骑兵攻击中证明了他们的威力。但到了公元4—5世纪，随着西罗马帝国军事力量的衰落，罗马鼎盛时期严整的步兵战阵，在欧洲战场变得松松垮垮，骑兵在战斗中又再度崛起。在第一联合作战范式中，剑、矛、弓箭、盾和盔甲依然保留着，但战场的中心位置易主。罗马帝国的最后几个世纪，以及中世纪最初的几个世纪里，步兵和骑兵的轮回再度启动。骑兵变得威力更强，而步兵退到了从属的地位。同二轮战车一样，变化的推力与其说是来自骑兵作战技术的增强，不如说是由于步兵自身纪律和训练松弛所导致的。

中世纪作战

公元5世纪，一个可以与公元前1200年前后的“大灾变”相提并论的黑暗时代，降临到了欧洲。罗马的税收和管理都垮了，政府权威、军事力量、经济体系、政治组织也每况愈下。第一联合作战范式依然存续，但由步兵到骑兵的切换却步履缓慢。自5世纪到14世纪，新的马上武器系统在慢慢形成，它包括三部分：第一，到中世纪的末期，骑士的盔甲已经从古典时代末期的锁子甲（由铁环连缀而成），演变成“百年战争”时期的连片甲，最后到16世纪又变成能防护马匹和骑手的整体甲；第二，这种沉重的盔甲重量落在马的身上，导致了自11世纪到13世纪开始的马匹选种繁育，到14和15世纪，迎来了历史学家R.H.C.戴维斯所称的“骏马时代”。这一时期伴随着马匹选育，马的食谱里开始加入燕麦和其他谷物。这种改变在军队后勤上意义重大。欧亚大草原上的轻装骑射手，全靠草来喂养他们体型较小的马匹，这给了他们巨大的机动性和近乎无限的活动范围；相比较之下，西方的重装骑士则脱离不了他们的军械库和补给车队。

第三个技术创新是中世纪骑士留下来的马镫。这个看似简单的设备，其实是一件技术制品。它在公元7世纪从亚洲传至东欧，到8世纪时传到西欧。研究中世纪的历史学家小林恩·怀特认为，马镫在西欧的出现，完美地补足了一直以来关于西方封建制度起源的理论缺憾。德国历史学家海因里希·布鲁纳在1887年提出，封建制本质上是一种建立在军事关系上的社会和政治制度。一块地盘上的国王或领主，将土地分给封臣（他们也可能把分到的土地继续分给下一层封臣），封臣则利用封地获得的收入来购买成为骑士所必需的昂贵武器和装备。为回报获得的封地及其产生的收入，封臣们对领主宣誓效忠，并承诺每年服四十天左右的兵役。但怀疑此理论的人会问：为什么欧洲封建制开始于8世纪初？怀特的解释是，因为那时正是马镫首次在西方出现的时间。它的出现，才使得武器和盔甲都很沉重的骑士成长为欧洲战场上的主导力量。马镫使重装骑士成为冲击力极强的武器——他们依仗手中的长矛，能够在敌方步兵和骑兵队伍里所向披靡。领主们给予骑士土地，土地带来的收入让受封者负担得起昂贵的装备和随从，然后骑士以服兵役作为回报。本质上，这种独创的、前所未有的骑士制是封建制的核心，政治、军事、经济、社会和司法通过它得到完美集合。马镫让骑士在战场上所向无敌，而骑士则强化了封建制。

31

对布鲁纳和怀特文章的批评，在过去半个世纪里占据了主导地位。反对的意见认为：重装骑士在马镫出现前就已经是主战力量；而土地分封制度，在公元732年普瓦捷战役后，法兰克国王查尔斯·马特尔想出这个方案之前，就早已普遍施行；欧洲的封建制并非“马镫说”所认为的那种整齐划一的社会制度；另外，骑兵在冲击时，马鞍的作用比马镫更为重要。一些学者批评林恩·怀特犯了“技术决定论”的错误，以为是马镫造就了封建制。实际上，怀特曾明确地驳斥过这种观点，认为自己只是提出：马镫在西方的出现差不多与普瓦捷战事同时，它不过是在8世纪欧洲政治、军事、经济、社会、司法等各种因素叠加的基础上，为封建制在中世纪社会的骤然成形提供了最后的催化剂。围绕马镫作用的不同解释所产生的争议，恰好提醒我们：“技术决定论”的观点通常只是哗众取宠的辞藻，从来不是历史事实。历史学家们会批评他们的同行是“技术决定论者”，但真正令人敬重的历史学家不会去实践这样的理论。相反，严谨的历史学家会从具体情境中去理解事件，利用各类分析寻找有解释力的答案。马镫是军民两用技术，有助于解释封建制，但并不能造就它。

尽管在史学界存在争议，但不可否认的是：自公元8世纪早期至12世纪结束，重装骑士在长达五百多年的时间里控制了欧洲战场。其成功的原因，与其说是其战斗力让人无法抗拒，不如说是对乌合之众的步兵所造成的心理冲击。罗马帝国崩溃后，他们对阵的是装备不良、组织不力、拼凑而成的步兵。换句话说，骑兵此番重出江湖，与以前的二轮战车如出一辙，都是利用了步兵的混乱和恐惧，所以能在五百多年的时间里碾压步兵。

但是，在13和14世纪，欧洲的重装骑士遭遇过几次重大的逆转，直到最后彻底地被火药拉下马来。第一次逆转，发生在成吉思汗（1162？—1227）把蒙古部落联盟整合成一支强大的帝国军队之后。成吉思汗的军队，征服了中国北部和直到里海的欧亚中部地区。他的儿子和继任者继续将蒙古的征服范围扩大到俄国，经欧亚草原打到了今天的布达佩斯，直接威胁到欧洲各封建军队保护的地盘。这支蒙古军队，虽然全部是由野蛮人组成，但在作战的各个维度上都高出欧洲人一筹。他们有自己的情报部门、复杂的通信系统，后勤粮草供给足以保障他们需求不多的人马，组成轻骑兵的战士自成年就在马背上使用复合反曲弓娴熟射猎，对敌策略既有分头出击的战略又有战术合围，战争伦理无情、嗜血、残暴，军队领导亲自征战、指挥得当。不仅如此，蒙古军队当时还让欧洲人首次见识了最具革命性的军事技术——火药。1241年，这些来自欧亚大草原的侵略者，横扫西方的重装骑士，从太平洋到大西洋，把他们的帝国扩成了人类历史上最大的版图。但随后在1242年，他们突然掉头，撤回了蒙古。这次挽救了西方文明的转向，并不是欧洲人的抵抗造成的，事实上是因为成吉思汗病故，所有部落必须回去会盟，以选出他的继任者。在这之后，蒙古人对欧洲的进攻不如先前有效。部分原因是因为欧洲加固了城防。不管怎样，蒙古人在1241年的进攻，让盛名之下的欧洲封建军队显得千疮百孔。

在中世纪的盛期，欧洲骑士遇到的挑战，并不仅仅是外来的入侵。英法“百年战争”（1337—1453）中，就有英国军队与法国33军队的对阵。两支军队的重心都在重装骑士，区别在于：英国骑士独有一支辅助力量——长弓兵。这些轻装的披甲步兵，操的是一张性能卓越的长弓——长度可达6—7英尺，而当时英国人的平均身高可能只有5—6英尺。使用这种紫杉长弓需要过人的力气，以及特别的搭弦、开弓和精准射击的技巧。这种弓产生的力量可达到100磅，足以射倒一匹马，或者穿透钢制护胸甲以外的其他护具。不仅如此，它还可以根据战场需要，快速发射。为防止敌方骑兵的强势冲击，15和16世纪的长弓兵会在阵地前安放尖头桩，作为路障来拒止骑兵。“百年战争”中，这些长弓兵在战斗中一再证明了英国军队与法国军队的不同之处，特别是在克雷西（1346）、普瓦捷（1356）和阿金库尔（1415）等地的战役中。法国人多次失败，是由于他们正面冲击长弓兵，没有弓箭掩护且战术混乱。正是有了长弓兵，才导致了两军的差异，使得英国人能够以少胜多，肆意游荡在法国的乡村。

14和15世纪，欧洲大陆的封建军队在入侵瑞士时，遭遇了类似的逆境。那些训练有素的民兵结成紧密的方阵，从各个方向刺出密密麻麻的长矛，迎击冲击的骑兵。通常情况下，战马在被长矛刺中之前，就会止住脚步。采用这种战术，需要士兵的勇气和坚守阵地的决心。一旦敌方骑兵的冲锋势头衰竭，那些手持战戟、斧锤长矛、流星锤等各种长柄武器的民兵就蜂拥而上，围攻陷入困顿的骑士。他们武器上的钩子能把敌人从马上钩下来。一旦落马，骑士身着的板甲也难保其不任人宰割：刀子可以插进头盔眼部的空隙，斧头也可以砍向脆弱的关节。带有斧刃的长柄武器能砍断马腿，使骑士倒地——要么死亡，要么被俘。有一些战斗，封建军队凭借人数优势，打败了瑞士的长矛兵；但更多的对决结果，是瑞士民兵占了上风。当封建军队的骑士最终走向没落，瑞士的一些长矛兵开始借助战场上的威名，出售起他们的服务，变成了雇佣兵——其中最为著名的是罗马教皇的侍卫，到今天依然如此。其他皇家卫士多佩剑以显示其历史传承，而教皇的瑞士侍卫手持的还是当年让人闻风丧胆的战戟。

在两百年的时间里，欧洲的骑士三度被反制的技术打败，最后只好在16世纪再一次将舞台中心出让给步兵。最近的这次轮回为什么花了这么久的时间？在众多的答案中，有两点直接与欧洲中世纪鼎盛时期的作战技术有关。第一点，封建制是军事、政治、经济、文化和社会权力的集合体，这赋予它巨大的制度惯性。骑士是这个制度的核心，他们可以撬动多重权力杠杆。第二点，一旦战事不利，骑士可以退归自己的城堡，中止对领主的义务，击退其他骑士的挑战，甚至抵抗蒙古人的攻击。中世纪时的攻城技术，较之先前古典时期并无多少改进，虽然可以攻破一些城池，但面对设计优良、防守得当的城堡往往以失败告终。并且很多分封协定，只要求封臣每年服四十天兵役，时间并不足以实施围城，所以中世纪的围困战大多半途而废。欧洲中世纪的骑士也许赢不了所有的战斗，但他终究可以退守自己固若金汤的城堡。

火药革命

火药的破坏力巨大，不仅是对单个的骑士而言，对于整个封建秩序，也是如此。正如英国的长弓兵、瑞士的长矛兵以及蒙古的骑兵当年各自使用的武器系统一样，火器开始在战场上大行其道。就连骑士退避的城堡，也被火炮轰开了口子。这些城堡的护墙一般比较高，但是墙体单薄——这也反证了自亚述王朝以来攻城技术的停滞。高城墙，是为了阻止云梯，但它们建得不够厚。只要攻城的火炮架设起来，很容易被撕开口子，让步兵长驱直入。领主们用火炮迫使他们的封臣就范，剥夺他们享有的军事权力，把他们的服兵役义务转换成了纳税义务。领主们用收取的税养自己的步兵，买更多的火炮，再制服更多的贵族骑士。随着权力不断集中，封建制让位给了君主制。历史学家克利福德·罗杰斯认为，这是西方历史中最重大的军事革命之一。

欧洲的这种政治和军事变化，只是火药的出现所带来的诸多变革之一。火药，是本书认为的三次伟大的军事革命中的第二次，它也被认为是自古以来最重要的发明之一。火药革命，至少引发了另外八个重大结果。第一，不论在军事作战领域，还是一般社会领域，它宣告了“化学能”时代——我称之为“碳时代”——的到来。火炮算是最早出现的内燃机，同后来出现的内燃机一样，它的能量来自碳基燃料，包括木头（或木炭，火药的配方之一）和化石燃料（煤、油和天然气）。它反衬了第一联合作战范式的技术天花板，即能量主要来自肌体的力量（也有少部分来自风力）。自此以后，军事作战的规模将突破至化学能的领域。一旦利用化学反应（比如火）的武器等军事技术被人类掌

36

握，扩大死伤的技术创新就遍地开花，世界也随之快速改变。在公元二千纪的剩余时间里，战争所导致的破坏程度和人类死亡数量突破了人类的想象。虽然就死亡率而言，史前战争和第一联合作战范式下的战争死亡率更高，但其中大多数死亡是由战争引发的疾病和饥荒导致的。化学能在世间释放出的杀伤力，在第二次世界大战期间，随着战场上的“钢铁风暴”和对德国德累斯顿、日本东京这样的人类文明城市进行的轰炸，达到了顶峰。

第二，火药改变了要塞的布防。古代和古典时期的攻城机械，都谈不上高效甚至有效。在欧洲，城墙变得高了但也变薄了。即便是君士坦丁堡无与伦比的城墙，也在新的火器攻击下轰然倒塌，导致了1453年城市的陷落。随着火炮威力的增长，老式城墙愈发不堪一击。要塞的布防，要么改变，要么面临失效。于是，到了中世纪的末期，意大利北部城邦开始尝试一种新的被称为“意式要塞”（trace italienne）的布防形式，再度开启攻防技术相互克制的新一轮竞争——这种始自新亚述人的竞争一直延续到20世纪。

第三，相较于刺击、砍杀和棒击，投射武器变得更为致命。虽然在火器出现之前，弓箭比其他武器杀伤敌人更多，但它仍然只是在“打了就跑”的战术中才用到的工具。古希腊人鄙弃这种野蛮人热衷的武器，在古典时期和中世纪的作战中，它发挥的只是次要的辅助作用。但到了此时，世界各地战场上主要的杀伤，都是由火枪或大炮隔着一定距离造成的。这种改变，无论是对西班牙作家米格尔·塞万提斯本人（他在参加勒班陀海战中身负重伤）还是他书中的虚拟人物堂吉诃德来说，都是令人大为震惊的。力气和武艺，敌不过扳机的轻轻一扣；勇气和荣誉，敌

37

不过凌空而至的死亡。

第四，火药将重装骑士拉下王座，将枪炮手送上了高位。新的轮回里，中世纪的骑兵让位给了步兵。正如堂吉诃德所恐惧的，火药，让那些肮脏、拙劣的平民手里平添了一件可以杀伤高贵骑士的工具。军事力量的这种变化，动摇的不仅是骑士阶层，更是整个社会——平民借机上位，而贵族岌岌可危。骑兵直到20世纪才彻底消失，但此刻已开始沦为战场上的配角，而先前担任这种骚扰角色的，都是缺乏男子气概的投石兵、弓箭手和部队前卫。十字弓曾让人闻风丧胆，但也没有产生如此巨大的效果。

第五，自公元前12世纪的“大灾变”以来在野战中一直占据中心地位的作战模式被取代，出现了第二联合作战范式。新的范式在步兵和骑兵这两个传统的战场主角外，加入了炮兵。从

图4 这幅图描述了1514年的奥尔沙战役中，波兰军队和承包商在浮桥上用人力搬动一门早期火炮的场景。也许正是这些武器让波兰–立陶宛联军意外获胜，赶走了人数占优的俄国军队

38

17世纪到第二次世界大战结束，战场指挥官们都是在这三支作战力量之间权衡取舍。步兵、骑兵和炮兵，这三者都被赋予了化学能，由此带来的火力充斥了整个战场。

第六，火力所需的弹药供给，为部队带来更重的后勤负担——这比喂养重装骑兵的马匹要难得多。部队穿越田野也许还能找到燕麦和其他谷物，现在除了防守严密的军械所和弹药库之外，基本找不到武器配件和弹药。不仅如此，拖在队伍尾部的辎重还容易成为敌人袭扰的目标。

第七，自从人类社群有了文明、游牧、野蛮的区分之后，文明国度第一次彻底消除了野蛮人构成的威胁。此前的人类历史上，来自欧亚大草原或北非沙漠的所谓野蛮战士，不止一次征服了文明国度。波斯、罗马、拜占庭、哈拉帕、中国都曾屈服过，甚至整个西方文明在1242年都差点被野蛮人征服。但此后不会再发生了。在火药革命之后，野蛮人也许还能抗拒文明国度的入侵，甚至会用火器对付来自文明国家的敌人，但由于缺少工业技术和体系，他们永远无法生产出自己的武器弹药。缺少这些，他们再也不能征服一个已经建立了枪炮生产体系的文明国家了。这种技术上的不对称，诱使众多的西方国家投入帝国扩张的冒险，这带来了“西方的崛起”。尽管这些冒险很多结局不佳，但门口的野蛮人再也不会对其构成亡国灭族的威胁了。

第八，火药对于海战的改变同样巨大，这个话题后面我们将会讨论。第九，也是最后一点，火药的运用只是一场影响巨大的两阶段技术革命的第一阶段。火药通过碳化合物爆炸释放出化学能，推动炮膛中的炮弹、枪管中的子弹以及炸弹的弹片高速飞出。第二波的碳燃烧技术在19世纪进一步扩展，碳化合物产生的化学能被用来驱动战争机器。在20世纪的世界大战中，这些机器将杀人和破坏的能力推到了新的高度。这第二阶段堪称“技术革命中的革命”又可以分为两个时期，第一个时期是19世纪，第二个也更为重要的时期是20世纪。碳时代的后半段将留待以后探讨。

在谈及其他的作战领域之前，一个值得回答的问题是：为什么发明火药的中国，没能发挥出它的潜力？而作为引进者的西方，却让它发挥了如此巨大的作用？历史学家威廉·H.麦克尼尔认为，这只是因为西方人好战而已。肯尼斯·蔡斯对此持不同意见。他认为，早期的火器太过笨重，难以有效地对付游牧民族。那些游牧民族来自他称之为“不毛之地”的欧亚大草原和北非沙漠，也就是我所指的“野蛮人”。火药技术，对于那些步兵即将重新成为主流的国家来说，比如西欧、日本和奥斯曼帝国，却是非常适宜。他认为，正是火药推动了步兵对骑兵的更替。罗伯特·奥康奈尔则认为，是工匠和资本家带来了西方在火药技术上的领先。不可否认的是，西方在科学和火药技术上多点开花，后来在各个方面都有了创新。毕竟，西方的文化是把自然视为征服的对象。

40

随着第二联合作战范式开始出现，我们对陆战的讨论将暂告一段落。火药革命横扫欧洲的各个阶段大约以百年为界。14世纪，火器出现；15世纪，火炮被用来摧毁城墙；16世纪，轻武器推动步兵重新上位，将重装骑兵拉下王座；17世纪，移动火炮为野战战场注入了第三支力量。到了20世纪上半叶，新的作战范式直接导致了总体战。

41

第三章

海战、空战、太空战和近现代作战

海　战

穿过古代历史的迷雾，我们看到在公元前二千纪呈现出的海战，是以桨帆船（桨为主力、帆为辅助的单层甲板船）实施作战的。不同于大多数陆战的是，海战受限于承载水兵和武器的平台，这一点同空战和太空战一样。海战通过三类平台实施，以其推动方式进行划分，分别为桨、帆和蒸汽，每一种都有各自的技术特点和作战方式。平台运用到的技术，必须能够把水兵和他们的武器载运到条件恶劣的海上进行作战，水兵的作战方式也由技术所限定。武器的攻击目标可以是敌船，也可以是敌方船员，但通常需要平台技术对作战技术进行补充。陆战中的二轮战车，甚至重装骑士，也是类似的平台和武器的结合体。在20世纪以前，舰船体现了最为复杂的军事技术，堪称“整合各种系统的系统”。

应商业发展的召唤，海军应运而生——既是为了攻击商船，也是为了保护商船，并且商业利益从此成为海上宣战的主要理由。在军舰出现以前，都是靠民用船艇跨越海洋，运输人员和货物。地中海是早期海战的试验场，也是军舰演化的档案馆。那里最初发展出一种特别的商船，后来演化成了军舰。这些船采用框架建造法，也就是先建造船体，然后再添加船筋等其他固定结构。这种方法造出的船，在相对平静的地中海行驶不是问题。但是，这种船的船体外面仅仅凭借一根龙骨来加固，并不牢靠。船体轻，使得地中海的这种桨帆船行驶速度快，但是不堪一击。 42

毫无疑问，海盗导致了军舰的兴起。行驶缓慢、没有武装的商船，非常容易受到航速更快的海盗船攻击。海盗们追上满载货物的船只，铆住船舷，然后登船抢劫。总之，海盗可以打了就跑。如果往商船上派兵，只会让船只速度更慢，而且增加了运输成本，却并不能保证船上的兵力足以抗击那些海上的掠食者。因此，到了公元前8世纪或公元前9世纪，一些海洋国家（如亚述、腓尼基等国家）开始建造特定用途的舰船，用来保护他们自己的船队——也许顺带可以打劫别人的商船。很快，这些特定用途的舰船就显出了与众不同的特征。为提升速度，他们加长了船体、减少了吃水量、增加了桨手，并把货船的圆形船头改成了尖的。采用的战术是用尖的冲角猛烈撞击敌人的民船或军舰，然后再向后划，留下被刺破的敌船瘫痪在那里。不久之后，这种尖状、金属包裹的冲角（罗马人称之为“鸟喙”）让位给了一种形状更钝的船头，目的是让它给敌船造成塌陷而不是刺穿。因为尖的船头可能会插入行将沉没的敌船，不仅使得攻击船无法动弹不说，还可能会让敌方士兵趁机翻过船舷占领己方船只。

公元前500年至公元500年，随着地中海各国开始打造海军舰队，军备竞赛出现了。在这场竞赛中，获得胜利的关键因素是 43 船速，而地中海桨帆船的构造，决定了只有一种方法来实现这个目标，那就是增加桨手。战船越造越长，最后承载的桨手达到了五十名——每侧有二十五人。在此之后，船只加长的速度减慢了，可能是因为制作龙骨的高大树木太过稀少的缘故。取而代之的是，桨手们被分层安置（希腊人的“多桨船”）。腓尼基人和亚述人使用双层桨船。雅典人造的多层桨船最为出色，有三层——这样原先一名桨手的位置可以上下排列三人。后来的海洋国家，如迦太基和罗马，都沿着雅典人的路径发扬光大，造了四层乃至五层战船。这些巨型战船的桨手位置如何排列，还存在很多疑问。不少学者认为，这些更大的数字应该是指增加了

图5　图为20世纪80年代仿制的古希腊三层桨帆船“奥林匹亚斯”号，1990年在希腊托隆港下水。作为武器平台的这种桨帆船，设计上是运用水下的船首冲角来撞击敌船。但更多的情形是，在敌船瘫痪后，士兵登上敌船进行肉搏

44

划每支桨的桨手数目，而不是增加了更多的船桨。

无论动力如何编排，这些桨动力战船无疑清楚地展示了船只巨型化的好处和弊端。如果一件武器是有效的，那么看起来尺寸更大的似乎会更有效。这一点在桨帆船上得到了某种程度的证明。更大的桨帆船会更结实，可以承载更多的桨手，甚至还可以架设对付港口防御工事的攻城机械。在对阵较小的桨帆船时，也不那么容易被对手刺破。多层的大型桨帆船，在吃水线上的干舷空间更大，士兵们居高临下，可以更容易地向甲板上的敌人投射武器或跳上敌船。当然，巨型战船有时候也会被更为敏捷的小船所包围；如果海盗或小船逃入浅水区，追击的大型船也会无能为力。但不管怎样，桨帆船依然是伟大的杰作，是那个时代最为复杂的移动技术系统，对所有试图争夺这片海域控制权的竞争者都起到了震慑作用。

随着船只的演化，它们自身的特点也决定了作战技术的演化。撞击敌船是理想的目标，但往往难以实现。常见的情形是：老练水手驾驶的战船，可以快速灵巧地避开敌人的船头，并在敌船收桨入舱之前，用船头破坏其一侧的船桨。然后，敌船不得不重新分配船桨。但在此之前，敌船会失去动力，陷入被撞击、登船或超越的被动境地。但这种战术在狭窄的水域无法有效施展，所以海战最终变成双方舰队并排展开的大规模缠斗。投射武器和登船战斗因此就成了决定性的因素。罗马人在“第一次布匿战争”（公元前264—前261）中与迦太基人发生海上冲突之前，主要进行陆战，在船只和水手方面殊为不利。为应付这种不对称的局面，他们借鉴地中海东部海战中的经验，采用了“乌鸦吊桥”——一种可以从船头伸出去的舷梯。这种装置挨着前桅杆，垂直立于支点，一旦敌船进入范围就搭扣到敌船的甲板上。随后罗马士兵——罗马的核心军事力量——就可以沿着舷梯冲向敌人，展开肉搏。简而言之，罗马人把海战变成了漂浮平台上的陆战。其中最为著名的，是发生在米拉和埃克诺穆斯的关键战役。在经历了自己的桨帆船舰队三起三落之后，罗马人终于在海上打败了迦太基人，夺取了后来的军事理论家阿尔弗雷德·泰勒·马汉所称的“制海权”。

但同其他的海洋国家一样，罗马人发现，制海权非常费钱。一艘桨帆船只能用二十至二十五年，还需要大量的补给和人力。几个海上强权，比如雅典和迦太基，也许有资金建立和维持常备海军（雅典把一座富裕银矿的大部分收入用来养舰队），但对罗马这样的陆上强权来说，同时维持守卫帝国的陆军和守卫地中海的海军，则疲态尽显——尤其是在并无真正的海上威胁的情况下。随后的几个世纪里，各国也都在陆地和海洋权力平衡的问题上艰难取舍。

罗马帝国衰落之后，地中海统一的制海权分崩离析，成为各国争夺的目标。其中，拜占庭最为接近实现控制地中海诸多区域的目标。他们在海上击退了帝国的挑战者，凭借的是一件真正意义上的古代秘密武器。所谓的“希腊火”是一种易燃物，很多特点与现代的燃烧弹相似。它被装在拜占庭的快帆船上，先在甲板下面预热，然后利用气压从船头的导管射出，导管口的火苗会点燃射出的液体。据说，它能粘住任何碰到的东西，即便在水下也能燃烧。公元677年，面对穆斯林对君士坦丁堡的围攻，靠它赶走了敌人的战船，让城市和帝国转危为安。这种燃烧物仅仅被用来保卫君士坦丁堡，它的配方一直被皇公贵族严格保密，直到最后随着宫廷内斗、拜占庭统治的终结而消失。当时的人以及后来者，都没能重现这种武器的威力。就算关于它的说法是被夸大的，还是有很多人愿意相信它的存在——这也使得它成为一种无与伦比的恐怖武器。

在桨帆船作战时期，自始至终都非常看重战船的巨型化。尽管罗马人的利波尼亚式战船和拜占庭的快帆船有别于此，但到了16世纪，被称为“加利亚索蒂尔”的威尼斯战船，拥有罗马人的五层桨帆船的尺寸，排水量却达到了它的两倍。1571年，地中海历史上最后一场桨帆船大战在勒班陀展开，参战的四支海军都努力地往船上加装大炮。信奉基督教的舰队平均每艘船装有五门炮。被他们击败的穆斯林舰队，规模更大，但每艘船平均不到三门炮。所有的参战者，都没能找到在桨帆船上加装更多舰载大炮的好办法；而桨帆船也因为在新的海战环境下适应性不佳，最终走向了灭亡。不光改变了陆战方式，也改变了整个世界历史的火药革命，在改变海战方式上的潜力同样巨大，但这需要一种不同于以往的作战平台。勒班陀海战中，最大的桨帆船可以沿中线正向安装一门大炮，船上其他的火器都是只能用来杀伤人员的轻武器。但是，对人员的杀伤，并不足以左右战局。大炮的威力则可以摧毁敌舰，而不仅仅是敌方人员。在交战中，平台的作用比人员更重要——虽然这一点可能不为美国独立战争期间的美国著名海军将领约翰·保罗·琼斯所认同。

近代早期（1500—1800）的两项重大技术革新，和几项次要的技术革新叠加在一起，使西方的舷炮战船得以出现——这是当时最为复杂的技术产品。首先，火药革命带来了一系列的武器。这些武器既可以用于陆地，也可以用于海上。其次，在中世纪晚期，航行于北大西洋的柯克船演化成为作战平台。这个过程同早期的桨帆船一样，先变成武装商船，后来进一步变成专为海战而造的军舰。柯克船形状宽扁，航行慢而稳，自中世纪早期以来就在波罗的海和北海，并沿着欧洲的大西洋海岸往返运输物资和人员。随着14世纪欧洲商业的勃兴，这片海域的贸易兴旺起来，因此招来了海盗。海盗们采用的是类似船只，装备有远程的投射武器和登船后的肉搏武器。商船于是同样地加强武装，同海盗船展开了小型的军备竞赛。为了在交战中占据优势，双方都在甲板上搭建了“塔楼”——弓箭手们在上面居高临下，射击敌船甲板上的人员。到了15世纪，单兵火器被用在塔楼上，很快大炮也被装到了船上。

但是，这个进程遭遇了技术上的天花板。塔楼位于吃水线上方，再在塔楼上架设沉重的大炮，会使得小船重心不稳。大炮开火的时候，后坐力也容易让船只倾覆。因此，船上只能装用来杀伤人员的武器。这个天花板后来被打破，源自一个意外的创新。商船的运货人，为了上下货物方便，在船的侧舷开设了舱门。这些舱门具有水密性，在航行中处于关闭状态，即便船身侧倾也不会进水。人们由此自然想到，这种舱门也可以被用来发射炮弹。大炮一旦从主甲板上方的塔楼移到底层甲板，那么限制一艘船装载火力的唯一因素，就是船的大小了。于是，一场新的追求舰船巨型化的竞赛展开了。12世纪的时候，上百吨重的柯克船的前后塔楼上只有几个弓箭手，而到了1700年，出现了排水量近两千吨的、装有上百门大炮的战船。柯克船主要以敌方人员作为攻击目标，而这些浮动的炮台，瞄准的是敌方舰船。

但是，如果不是有了另外的几项创新，装载了舷炮的战船威力也无法得到充分发挥。以前的桨帆船，一直用桨来控制航向，到了12世纪末，艉柱的舵取代了桨。这种舵连接到甲板上的舵轮或舵柄，对大型航船来说不可或缺。但是，遇到天气不好，舵手还是很难掌握航向。指南针，在12世纪末、13世纪初在欧洲出现。有了它，水手们可以更远地离开海岸，并最终敢于向大西洋深处探索。在18世纪，先进的六分仪，取代了之前沿用了几百年的落后的天体观测仪和十字测天仪。这让航海者在看不见陆地的情况下依然可以估算出纬度，即赤道以南或以北的距离。最后，同样是在18世纪，英国的约翰·哈里森做出了精致的航海天文钟，即便是在颠簸的海上，它依然可以非常准确地计时。有了它，水手们能够确定经度，即他们的位置在某个特定地点的东或西。纬度加上经度，就可以让航海者在茫茫大海中找到准确的定位。

装上了舷炮的航船拥有了前所未有的能力。因为动力来自风——一种生生不息的能源，那么能限制船只活动范围的因素，就只有船员的食物和淡水补给了。由于这些东西在世界各地都能获得，于是这种航船的活动可以变得无边无界。在遨游世界各大洋之际，任何海上的其他类型船只都无法对它构成威胁——无论是地中海的桨帆船，还是中国的平底帆船，或是南亚的独桅帆船。实际上，这些战船的威力如此巨大，以至于它们在公海遭遇的竞争者只会是同类。有意争霸海上的国家，为此开展军备竞赛，所造的战船尺寸越来越惊人。各国的军力在此过程中拉开了差距。构成舰队的主力是“战列舰”。这些船上至少装有六门大炮，体型和火力足以击败当时海上最大的船只。这些海上堡垒，造价高昂，只有最富裕的国家才能负担得起。同二轮战车一样，战列舰的出现，迫使想要争夺制海权的国家面临这样的选择：要么参加军备竞赛，要么退出竞争。

这种竞争既带来无尽的好处，也带来巨大的风险。海上强国不仅可以保护自己的商船，掠夺敌人的商船——就像早前的桨帆船干过的一样，还可以将欧洲的海军力量投射到岸上。通过将资源聚集到海军建设，荷兰和英国这样的海洋国家迅速崛起。而像西班牙和法国这样的国家，走的还是罗马人的老路，既想做陆地大国，又想做海洋大国。它们最终没能成功——资金耗尽，军事上也陷入失败。近代早期围绕制海权和帝国地位的争夺，以1805年的特拉法加海战而告终。在这次海战中，帆船时代最伟大的海军将领霍雷肖·纳尔逊，带领英国皇家海军击败了法国和西班牙组成的联合舰队。虽然纳尔逊在战斗中负伤而亡，英国却由此成为无可辩驳的海上霸主，开启了一个"不列颠治下的和平时代"，并一直延续到第一次世界大战。

在这场帆船时代的巅峰之战中，纳尔逊使用的旗舰"胜利"号拥有一百门炮的火力系统、三千五百吨的排水量、八百多名的船员和战斗人员。它沿袭了北海地区的柯克船模式，船舷两侧开有舱门，便于上下货物。但是，"胜利"号和它同时代的战列舰都有技术天花板。这个天花板限制了它们的行动，并最终导致了战列舰的消亡。同它们所取代的桨帆船一样，战列舰极其昂贵。各国为了建造和维持它们，需要砍伐乡间的许多大树，还要到街头和小酒馆里去搜罗那些穷困潦倒的男人，让他们到船上去当水手和炮手。对风力的依赖，限制了船的航速。它们是可以去到任何地方，但是很慢；碰上逆风，只能迂回曲折，方能到达。

图6 1805年特拉法加决战中，霍雷肖·纳尔逊的旗舰“胜利”号成了悲喜交加的英国人夺取海上霸权的象征。爱尔兰艺术家丹尼尔·麦克利斯的这幅画，将“胜利”号绘作战场的重心，这里也是纳尔逊被狙击手射中之后理想的牺牲之地

不仅如此，当进入交战状态时，为了让炮口正面迎敌，他们还不得不调整航向姿态。也就是说，作为武器系统的船只，同时也是瞄准系统。因此他们一般保持一字纵向的队列航行，接敌时，整个舰队与敌人的舰队平行。这种战术让双方近距离猛烈交火，却难做到致命一击。实际上，纳尔逊之所以大胜对手，正是因为选对了攻击角度。

到特拉法加战役之时，帆船的时代已经式微。美国的艺术家兼工程师罗伯特·富尔顿，此前已经在巴黎造出了蒸汽轮船。在纳尔逊死后不到两年的时间里，富尔顿的第一艘商船“老北河”号也将启动。但无论是富尔顿的这些船，还是出现在美国海岸边和内河里的那些仿造者和竞争者，都不能对英国皇家海军的“胜利”号构成多大的威胁。“老北河”号的长度只有一百五十英尺，引擎的震动震耳欲聋，就算在平静的水面似乎也能把船体撕裂。早期的蒸汽轮船采用木制的桨轮，一旦遭遇炮击会立即分崩离析。船上如果加装炮位会更为不堪，到了公海，不等敌人开炮大概就已经解体了。但它们毕竟开启了一场技术革命。在其后一百年里，诞生了第一艘全重炮战舰——英国皇家海军的“无畏”号。在“无畏”号出现四十年后，日本“武藏”号战列舰在敌人的火力攻击下，带着舰上二千四百名船员葬身锡布延海海底——这只在“巨型化”竞赛中拔得头筹的巨兽，落了个徒劳无功的下场。

从“老北河”号到“无畏”号，在19世纪新技术层出不穷的那几十年里，这个故事的第一阶段得以展开。蒸汽船这种可以军民两用的技术，最初是为商业目的而研制，然后一直沿此轨道单一发展，直到后来出现专门的军舰。蒸汽机引擎，在使用了双向做功的活塞和高压蒸汽以后，功率大增。1819年，第一艘蒸汽轮船成功跨越大西洋。但海军的军官们却抗拒这种新技术，认为它不可靠，而且样子也太丑。就算蒸汽机登上了海军舰船，也只是用作辅助性的发动机，装在传统的舷炮风帆战列舰上——这种混搭让人想起了以前那些挂上了风帆的桨船。当然，在此期间，海军的火炮在不断改进；与之对应，传统的木质船体上开始尝试加装铁制装甲。1862年，两艘完全由蒸汽机驱动的装甲军舰在美国南北战争的战场相遇，其中一艘“莫尼特”号是专门设计的，全金属船体，旋转炮塔上装有两门炮，另一艘“梅里马克”号是木质船体外包覆铁甲，装有舷炮，这从此宣告了海战中蒸汽机时代的到来。自那以后，海军技术发展飞速：船体材料由铁变成钢，主炮塔装上了膛线火炮，船体和甲板上的装甲将战舰变成了海上堡垒，涡轮机让航速达到二十节甚至更高，无线电让舰船内部和舰队之间实现通信，陀螺仪让军舰和火炮更稳，测距仪让这些海上霸主打击范围扩大到十英里甚至更远。 52

与此同时，另一条碳基热能发动机的开发路线，改变了19世纪的海军军备竞赛，并引发了后来的世界大战中的政局动荡和无数毁伤。19世纪60年代，尼古拉斯·奥托率先研制出先进的四缸液体燃料内燃发动机——这是一项军民两用技术。这是有史以来第一次，碳化合物以液态化石燃料形式储存的能量，可以被人们通过发动机的气缸加以掌控。蒸汽船还是老样子，依然通过锅炉烧煤或油产生的水蒸气驱动。但真正的内燃机的出现，意味着只要有台紧凑的燃油发动机，就可以驱动尺寸不那么大的机器到处活动——下至水面，上至大气层顶端。内燃机的技术应用，包括了两种能够摧毁大型舰船的技术。

一种是潜艇。几个世纪以来，人们一直试图建造水下武器。1800年，蒸汽轮船的发明者罗伯特·富尔顿为拿破仑造了一艘，并带领三名船员航行到英吉利海峡，去袭击在那里执行封锁任务的英国海军，但没能成功。美国南北战争期间，南部邦联的"汉利"号潜艇成功击沉了联邦军队一艘军舰，尽管它自己最终也在行动中沉没。这些潜艇的驱动装置固然精巧，却是以人力转动曲轴，动力始终不足。直到内燃机和蓄电池的发明，才使潜艇适用于航海，且用处巨大。1897年，美国人约翰·菲利普·霍兰德造出了第一艘现代潜艇的原型机。十七年后的第一次世界大战中，对商船的袭击就来自威力巨大的潜艇。

内燃机还被装在飞机上，这是另一项军民两用技术。飞机的发明，最先是为民用，但很快便被军队采用。1908年，莱特兄 53

弟通过多场表演，展示了飞行的技术。几年之内，他们的发明便将人类冲突拓展到了第三维度，也改变了第一次世界大战的战场。在这同一场战争中，使用内燃机的坦克，宣告了骑兵在轮回中的再度回归，并在20世纪的大部分时间里主宰了陆战。

内燃机与蒸汽机——它在碳时代的前身——相比，其驱动的机器对于作战方式的改变更为全面，其中最为显著的改变体现在海战上。正如之前桨帆船发展为多层战船，再到战列舰的历程一样，蒸汽机战舰在其问世的头一百年里，也从美国内战时期著名的小型单炮塔军舰“莫尼特”号，发展成前所未有的巨型日本军舰“大和”号及其姊妹舰“武藏”号。这两艘军舰的满载排水量达7.28万长吨——是纳尔逊的“胜利”号的18倍。它们有9门口径超过18英寸的主炮，能将3200磅重的炮弹射到26英里之外——与纳尔逊最大的火炮相比，炮弹的重量和射程分别是前者的100倍和26倍。然而，最终击沉“大和”号的，却是重量不到其百分之一的舰载俯冲轰炸机和鱼雷飞机。

“巨型化”的冲动，再度引诱人们造出了巨型战舰。贪图蛮力，却忽视了灵活机动和精准打击的潜在力量，这在人类历史中反复出现。大和小，在各领域的作战中，其实难辨其优劣。确切地说，技术能放大大块头的蛮力，但也能增加小个子的灵活度。像内燃机这样的技术，既是调平器，也是倍增器。20世纪以来的战争，在很多方面表现为：为完成特定任务而寻找合适技术的竞争。质量和数量、大与小、新与旧，都难言孰优孰劣。碳时代的变迁，充分体现在海战上，技术推动引发了变革。也就是说，在别的地方研发的技术，改变了海战的形式。火药催生了舷炮战船，蒸汽机催生了蒸汽战舰，而内燃机驱动的飞机则干掉了巨无霸级的蒸汽战舰。

碳时代到核能时代的转换，对海战的改变依然显著。原子能或称核能，是比较少见的那种军民两用技术。它的发明者从一开始，就意识到了它的军事价值。在20世纪20和30年代，随着原子的秘密被迅速揭示出来，物理学家们看到了从某些重元素中分裂原子的可能性。如果能将原子分裂的话，这个过程将释放出天量的能量。如果一个原子分裂中释放的中子，继续分裂其他的原子，那么可能会产生链式反应。从理论上来说，链式反应可加以控制，让其缓慢或骤然释放能量——前者可用于发电，后者可用来制造炸弹。1938年，德国科学家奥托·哈恩和弗里茨·斯特拉斯曼，在流亡的同事丽斯·梅特纳的帮助之下，成功地用中子轰击并分裂了一颗原子。随后，全世界的科学家迅速将其投入应用。当时的欧洲，第二次世界大战迫在眉睫，于是各国展开了生产出第一颗原子弹的竞赛。美国实施的"曼哈顿项目"——也有来自英国和加拿大的科学家合作——最终在第二次世界大战期间领先对手，并于1945年在日本的广岛和长崎展示了他们的研究成果。

第二次世界大战末期，美国海军舰长海曼·里科弗比大多数人更清楚地看到了原子能（他称之为"核能"）在炸弹之外的军事用途。如果能很好地控制链式反应，核聚变也许能被用来为军舰提供动力，这样就能解决蒸汽时代以来困扰人们的两个问题：一是，不必像蒸汽船那样定期停靠码头补充煤或油，因为核动力船只可以多年不用补充能源；二是，由于核反应堆提供动力不需要氧气，因此造一艘真正的潜艇成为可能——这种潜艇可以在水下持续潜航几周甚至几个月。里科弗说服了美国海军部，允许他对这些可能性加以尝试。

在曾参与“曼哈顿项目”的橡树岭国家实验室，里科弗学习了有关核反应堆的技术知识。随后，他获得批准去开展船用核能发电机的试点项目，首先是为潜艇找到一种核动力装置。刚刚起步，他就遇到了技术方面的几个关键问题。他需要选择一种燃料，一种能放缓中子释放速度的减速剂，一种能维持核心稳定的冷却剂，一种热交换器来把核反应堆产生的能量转化成推动涡轮的水蒸气，一种能将中子反射回反应堆以防止辐射泄漏的覆层材料，最后还需要能加速或关闭链式反应的控制棒。考虑到潜水艇内空间的局限性，里科弗选择了轻水反应堆。这种设计用加压的水来为反应堆冷却和减速，另外还有一套水系统提供能量转化。他为航空母舰设计的是类似的但尺寸更大的沸水反应堆。

正是这台在宾夕法尼亚州希平港建造的原型轻水反应堆，产生了最大的影响。这个为核潜艇提供动力的装置研发成功后，被装在美国海军“鹦鹉螺”号上，并于1955年进行了首航。从那以后，核反应堆为两类美国潜艇提供动力，即攻击潜艇和所谓的战略潜艇——能发射带核弹头弹道导弹的潜艇。当潜艇携带的导弹升级为洲际弹道导弹时，潜射弹道导弹就成为美国战略攻击三极（远程轰炸机、陆基导弹和潜射导弹）中最抗打击的一极。苏联的武器可以打击停在基地地面的远程轰炸机，也可以打击发射架上的洲际弹道导弹，但冷战期间的苏联一直不曾拥有找到并摧毁藏在大洋深处的美军战略潜艇的能力。因此，潜射弹道导弹成了冷战时期核威慑的终极手段。

56

里科弗主持的海军核反应堆项目所产生的影响，不仅仅局限在战略威慑方面。在项目早期，他就选定了轻水反应堆，并委托西屋电气公司建造第一座潜艇上使用的反应堆——这两点产生了瀑布效应。经济学家常常称之为“锁定效应”，指的是在文化或制度上选定一种技术发展路径，而略过其他的选项。研究技术发展史的社会学家称之为“闭合效应”，指的是一旦一种技术被选定，其他的技术就会退出舞台，相互之间竞争也告一段落。历史学家托马斯·休斯把这种现象称为“技术的动能”，特意与“技术决定论”区别开。因为后者是恼人的话题，在技术史研究中屡遭诟病。上述这些类比想表达的都是，研究群体在选定技术路径时并没有什么是“必然的”。为完成某项工作，很少的情况是只有一种“最佳的”选择。更真实的情况是，在不同的时点，不同的群体找到的是一种与他们的需求、资源和脾性更为对路的技术。不过，一旦他们表示出了对一种选择优于其他的偏好，他们也就为这项选择注入了动能。他们会放弃对其他选项的进一步研发，而且把整个研究群体锁定在他们选定的投入方向。

轻水反应堆就是这样的情况。在德怀特·D. 艾森豪威尔政府倡导的“和平原子能”计划的鼓励之下，西屋电气公司设计了一座轻水反应堆用于商业发电。当时围绕燃料、减速剂、冷却剂、覆层、控制棒等核反应堆要素，还有很多其他的技术组合方案，但是西屋电气只会选择自己已经掌握了的技术。按照这个发展路径，美国就此开启了它的商业用途——核电产业。经济学家把这种发展模式称之为“路径依赖”，指的是，一个领域发展到终点的路径并不是随意的，它受制于最初启用的路径。在这条路径上走得越长，再走回到其他未选路径的可能性就越小。

因此，踏上依赖路径的最初几步，就显得格外举足轻重——而里科弗就是在早期做出了那些关乎该技术命运发展的决定。到了20世纪70年代末期，由于多种原因的影响，美国对于核能的热度退潮，第二代核能技术的研发动力不足。其中一个原因在于，核能技术早期的选择是为了满足军事用途的需要而做出的。

当然，美国的第一代核能走入低谷的主要原因还是安全问题。1979年，三哩岛核电站事故为起步艰难但势头良好的核能产业敲响了丧钟。不过里科弗一直坚称，核能只要管理得当，就是安全的。1982年，当里科弗不再执掌海军的核能项目时，他特别强调的是，如他承诺的那样，没有一艘美军舰艇因核事故而沉没或严重受损。1963年沉没的"长尾鲨"号攻击型潜艇——所有乘员葬身大西洋海底，无一幸存，其原因在于机械故障，与潜艇上的核反应堆无直接关联。美国海军的核动力舰艇通过严格的教育、培训和纪律来避免事故的发生，这些工作都由里科弗亲自监督。他的整个职业生涯，充分证明了人的能动性所能发挥的作用，特别是人在安全使用那些似乎与危险相伴而生的技术中所能发挥的力量。正如政治学家兰登·温纳所说的，现代复杂的技术系统有时候似乎能"自动运行"，或者"脱离人的控制"。但里科弗却证明了，人类管理风险的能力，是可以成功地超越一般情况的。

空　战

飞机最初是利用两辆自行车的结构发明出来的，是一种军民两用技术。军方花了番工夫才想出来如何在军事上利用它。从1899年起，威尔伯·莱特和奥维尔·莱特兄弟俩就开始对当时所有关于人类飞行的资料进行了系统研究。后来，他们设计并测试了他们的翼型，开发了升降台来安装他们设计的机翼和推进器。为了能像鸟一样控制飞行姿态，他们发明了独创的机翼变形系统——通过滑轮和绳索改变机翼形状来实现不同的升力。他们先在地面用绳系住机身，在风中实验操控；然后把这架“大风筝”带到山顶，让它自由滑翔。这时，他们已经掌握了飞行技巧。为给飞机提供动力，他们委托一位机械师按照指定的规格设计制造发动机，并亲自设计了飞机的螺旋桨。1903年冬天，他们把所有配件组装到一起，凭借飞机自身动力飞行了852英尺。这样两位发明者，全凭一己之力，未获他人指导，仅仅通过阅读、思考、观察、理论化和实践，最后在如此短的时间里开发出一项影响巨大的新技术——可谓前无古人，后无来者。在等待专利申请获得通过的时间里，他们花了五年在他们的自行车铺旁边练习飞行。1908年，他们在巴黎和华盛顿展示了他们的成果，折服了在场的所有不带偏见的观众。他们全靠自己，最终解决了很多政府、机构和个人经年累月都未曾解决的难题。

世人又是如何接受飞行的这个礼物呢？一些人利用飞机来俯瞰世界：平民从上面拍照，军人则从上面侦察战场。美国陆军负责侦察的通信部队采购了第一批莱特飞行器，他们当时怎么也想象不出：这些由木棍和布组成的看起来弱不禁风的平台，将来有朝一日会被用来运载货物和乘客、大炮和炸弹。

但是很快，作为军备竞赛的一部分，欧洲的研究者开始竞相开发更快、机动性更强的飞机，这也加快了欧洲滑向第一次世界大战的进程。法国战场上空出现了新型的、更快的飞机，双方遭遇之后开始争夺后来所谓的“空中优势”。原先用于观察地面的飞机被改造成作战工具，上面加装了机关枪进行格斗——这让人们不禁回想起早年间，骑士们为了荣誉而进行的一对一的决斗。这种空战导致的伤亡和破坏当时仅限于飞行员彼此之间，所以没有引起世人的关注，更不会想到它在未来会产生的可怕影响。德国人曾尝试用两架体型巨大的多翼飞机（“哥达”号和“巨人”号）轰炸英国，但飞行员能做的，只是从开放的座舱把手雷和石块投向地面的敌人。所以第一次世界大战中的飞机，主要还是一种运输工具，一种可以用于观察和驱离对方观察者的平台。直到第一次世界大战之后、第二次世界大战之前，这种新的平台才真正找到了它的军事和民间用途，并最终改变了世界。

飞机在军事上的两种用途，自然呈现了出来。欧洲大陆上的国家把重点放在战斗机上，采用大功率的液冷直列式发动机以获得速度和机动性。他们追求的是战场上的制空权，以便实施侦察和攻击敌人的地面部队。美国和英国，则按照意大利军事理论家朱利奥·杜黑的制空理论，重点发展战略轰炸能力。这种达到远程的能力，需要完全不同的空中平台——一种由径向气冷式发动机推动的更大的机型。算不上巧合的是，这两个国家正好也需要拥有同样性能的商用飞机，这样航空公司才能把乘客运送至全美或幅员辽阔的英帝国各地。德国人从西班牙内战（1936—1939）中借鉴了经验，意识到了更为全面的制空权的重要性，于是开发了全系列的飞机以遂行各种任务：从空中格斗到中远程轰炸，再到空降兵突击。最终的情况是，英国人为了保卫本土，不得不在轰炸机之外补充战斗机；美国人也不得不增加战斗机，来护卫在敌方领土实施执行任务的轰炸机。英国和美国也都实验了用于海上作战的歼击机。与桨帆船和风帆船不同的是，作为作战平台的飞机，从一开始就是为装载特定武器和执行作战任务量身打造的。

第一次世界大战后、第二次世界大战前受到热捧的一款军民两用飞机在1935年上天，这就是DC-3——由美国道格拉斯航空公司设计的第三代商用客机。飞机采用了悬臂翼、带风冷罩的旋转发动机、变距螺旋桨、可伸缩的起落架、襟翼、流线型的抗压整体机身以及平齐铆接等技术，反映了当时最先进的工艺。道格拉斯公司制造了六百多架这种飞机用于商业运输，直到1942年转向开始专注于军用飞机，并于第二次世界大战期间生产了一万多架由它改装的军用运输机——C-47和C-53。早在20世纪30年代，道格拉斯公司就授权苏联和日本生产各自版本的这款飞机达五千多架，在第二次世界大战期间这些飞机也转向了军事用途。历史上还没有哪款飞机比得上DC-3的实用性和长寿命——它在有些地区今天依然在被使用。

民用DC-3机型长盛不衰，这与军用飞机按需定制、一路呵护的发展历程截然不同。两次世界大战间隔时期所举行的各种飞行比赛及其奖金，吸引了众多的参赛者，比如在1927年首次成功飞越大西洋的查尔斯·林德伯格，同时也刺激了技术的革新。随着军队需求增加，技术能力所拓展到的领域已远远超出当初制空权热衷者的想象。战斗机、护航机、侦察机、运输机和歼击机等机型纷纷出现，开始更好地在陆地和海洋上空执行任务。美国人在第二次世界大战初期使用的B-17轰炸机（飞行时速287英里，飞行高度3.5万英尺，飞行距离2000英里，载弹量6000磅），到战争末期轰炸日本广岛和长崎时已升级成巨无霸B-29

61 轰炸机（飞行时速357英里，飞行高度3.2万英尺，飞行距离3250英里，载弹量2万磅）。不过，这些军事上的重大发明所产生的效果，与早期看好飞机的那些人的预言不尽一致。战略轰炸机从未像他们预测的那样，发挥一击制胜的作用；其他空中力量的发挥也取决于陆地或海上具体的作战情境。比如，近地面的空中支援，比他们当初设想的更能左右地面战场形势；运输机也比轮船更能快速安全地把人员和物资运送到世界各地。飞机运载的空降兵大大提升了地面战的战略机动性，这是先前的骑兵部队望尘莫及的。

不仅如此，配套的技术也出现了，用以辅助或反制快速发展

图7　这张不甚清晰的动作照片捕捉了1945年5月美国B–29轰炸机向日本横滨投下燃烧弹的瞬间。这些“超级空中堡垒”能携带2万磅的炸弹飞行3250英里，达到350英里的最高时速和3万英尺的飞行高度——超出了战斗机所能达到的高度。两架外号分别为“艾诺拉·盖伊”和“博克斯卡”的这种机型，1945年8月在日本投下了原子弹。B–29是当时的终极武器

62 系统

的空中力量。1940年的不列颠空战中，沿着英国东海岸部署的长波雷达为英国的截击机提供了关键的预警。

短波雷达则是战争期间最为关键并且花样繁多的发明，它产生了一百多种用途，包括机载雷达和近炸引信。改进后的无线电通信让地面指挥能与空中支援力量直接联系；新的投弹瞄准器使美军所谓的“精准轰炸”成为可能——尽管其实战准头与试验场还存在差别；远距离无线电导航系统（LORAN）能引导轰炸机攻击远程目标；防空武器为地面应对空袭提供了防御；还有，在战争末期终于出场的原子武器，不仅让无论是盟军还是日本人都避免了两栖登陆日本的重大伤亡，也让制空权热衷者所鼓吹的空战决胜论找回了一些颜面。

第二次世界大战之后，雷达技术与计算机技术融合，形成了能自动做出防御反应的防空系统，同时也催生了军民两用的计算机网络技术。早期的无人飞行器实验为后来的无人驾驶航空器（也被称为“无人机”）打下了基础。空气动力学在飞机设计上的应用，如后掠机翼、蜂腰机身，催生了超音速飞机。火炮和导弹也被装在飞机上，在空战中一争高下。空中加油技术大大拓展了军机的活动范围。

最后，航空的发展带给人们一个体制模式问题，即，如何将军事技术的创新常规化和制度化？由于在空中比在陆地和海洋作战的技术挑战性更大，空军的技术淘汰周期比海军要短得多。一代机型尚在服役之时，它的替代机型就已经在研发中了。新的一代机型必须飞得更高、更快、更远，火力必须更准、更强、更猛，技术支持必须更安全、可靠和高效。历史学家所谓的“性能贪欲”在空军出现得比其他军种更早，这也把他们推入了花费高、强度大的尖端武器军备竞赛。在他们引领之下，其他的军种也跟随而来，很快就导致了德怀特·艾森豪威尔总统所谓的“军工联合体”的出现。

太空战

太空战的起源与航空在很多方面相仿：两者最初都是由受到灵感启发的非专业人士纯粹出于爱好所发明，并无将其用于军事的打算。但不久之后，该技术在军事上的用途就显现了出来。飞机和航天飞机很快就被用作服务军事行动的平台，这也正好体现了军民两用技术在需求拉动和技术推动下形成新应用的过程。与飞机不同的是，航天飞机并没有成为当初“太空战”宣扬者所设想的那种武器平台。

可以这样说，太空航行技术发轫于20世纪20年代，推动它发展的是一些理论研究者和目光深远的人，他们预言人类能够并且终将旅行到月球、火星，甚至更远的地方。把这样的幻想落到现实的，是两个毕生以此作为事业的杰出团队——德国科学家韦恩赫尔·冯·布劳恩领导的“航天研究会”以及美国科学家罗伯特·戈达德领导的规模小得多的团队。两者都使用早期的液体燃料小型火箭做了试验。在取得了一些成果之后，他们都寻求外部资金支持以建造昂贵得多的大型火箭，这样可以把巨大的载荷送至更高的高度。冯·布劳恩和他的同事求助于德国国防军。戈达德则从各种公私金主那里寻求支持——有史密森学会、美国海军，还有古根海姆家族。第二次世界大战前德国和美国不断加速的军事研发，也把火箭研究裹胁进来，偏离了冯·布劳恩和戈达德当初设想的民用航天而转向军事武器。战争结束之前，冯·布劳恩的团队已经研发出V-1火箭（V指“复仇”）——一种自动飞行的巡航导弹，后来又有了更出名、更致命的V-2弹道导弹，它能够携带1000公斤的弹头飞行200英里。第二次世界大战快结束的那几年，德国向同盟国的目标发射了3000多枚V-2，但由于制导系统的缺陷，德国人炸死的敌人还没有他们在自己的强制劳动营里致死的人多。 64

第二次世界大战结束时，美国和苏联俘获了德国V-2项目的大部分装备和人员，他们让这些人为自己的导弹开发项目工作。苏联人对远程导弹的需求更为迫切，因此在1947年就开始了雄心勃勃的洲际弹道导弹（ICBM）计划，希望能造出携带核弹头（此时他们还不精通此技术）并且能从苏联本土打到美国的武器。相比之下，美国是在知晓苏联的进展之后才奋起直追，开始了对洲际弹道导弹的研发。在这场对称武器系统的竞赛中先行胜出的是苏联，它在1957年10月4日向全世界展示了他们的成绩——这一天他们用新研发的洲际弹道导弹系统将民用科研人造地球卫星（Sputnik I）发射至轨道。尽管艾森豪威尔总统并不想将太空军事化，但混合了军事和民用目的的太空竞赛还是展开了。

太空竞赛在两条平行的轨道展开，两者都依赖洲际弹道导弹的军民两用技术。作为发射载体，它们将最初的人造卫星以及后来的宇航员送入太空；同时，军事版本的这种液体燃料火箭可以装载核弹头。它们与飞行员驾驶的战略轰炸机一起，构成了三极战略威慑体系中的两极。凭此战略威慑，美国和苏联打了一场从20世纪50年代一直到80年代的冷战。随着美国海军成功地将固态燃料弹道导弹装载到海军上将里科弗主持研发的 65

图8　1962年2月20日，水星–阿特拉斯6型火箭从卡纳维拉尔角空军基地首次将宇航员约翰·格伦乘坐的“友谊7”号太空舱发射至太空轨道。阿特拉斯系列火箭是洲际弹道导弹系统，也是太空运载工具，至今仍在军事和民用领域发挥作用

核动力潜艇，战略威慑体系的第三极也在20世纪60年代加入了进来。

美国人于1969年率先踏足月球，从而在平民版本的太空竞赛中获得领先。运载宇航员的是阿波罗飞船——一种特别定制的民用火箭，其背后的策划者正是改换了门庭的韦恩赫尔·冯·布劳恩。此时他已被美国国家航空航天局（NASA）所用，被赋予的任务是建造民用而非军用的太空运载工具。至此，这位一直向往飞向月球的航天迷终于摆脱军方束缚，得偿夙愿。

冯·布劳恩开创的军民两用航天技术，在第二次世界大战期间和冷战早期被军事需求所拉动，直到20世纪50年代才转向，推动了美国民用航天项目的发展。在把他的过人才华和勃勃雄心付诸民用航天的过程中，冯·布劳恩也给20世纪末期打上了历史学家所称的“布劳恩范式”的时代烙印。这种模式设想的是，先用液体燃料火箭（他开发的V-2的后继型号）将人和物资送至近地轨道；然后由宇航员在那里建立空间站，再以此为基地载人飞向月球、火星以及更远的敌方。即使到了21世纪的第二个十年，美国航空航天局依然以此模式指导他们的远期规划。

历史学家沃尔特·麦克杜格尔认为，美国的航空航天局及其民用太空项目实际上都是冷战的延续，只不过换了种方式而已。其对手苏联则把他们的军用太空行动整合到“火箭军”，成为第四军种，与陆军、海军和空军享有同样的地位。苏联的民用太空项目被分散到几个相互竞争的设计局，由中央政府统一管理。苏联的这种机构设置模式给人的感觉就是，太空一定会成为另一个作战领域——就像历史上战争从陆地延伸到海洋后

67

来又延伸到了天空，只不过在太空使用的武器平台更为复杂、昂贵，比地球上的任何地方都更危险而已。

但是，后来战火并未蔓延到太空。在20世纪50年代，美国总统艾森豪威尔顶住了国内“赤色恐慌”所带来的压力。当时，约瑟夫·麦卡锡及其追随者的危言耸听甚嚣尘上。他也没有听从“军工联合体”所谓“太空军事化不可避免”的警示论断。他的民主党接任者约翰·肯尼迪和林登·约翰逊，在就任后都更改了1960年大选中的高调，转而通过一系列的协议和政策，实施了艾森豪威尔放缓太空军事化的谨慎策略。

1967年，随着《外层空间条约》签署，两个超级大国和大多数工业国都同意：不将大规模杀伤性武器放入太空，不对地外星体主张国家权利，不干涉其他国家的太空轨道平台。太空航行的高成本、飞船轨道运行的危险，以及使用绕地飞船作为攻击平台的困难，让所有人都认为太空武器是个很糟糕的主意。

一直到1983年，罗纳德·里根总统提出了他部分基于太空的战略防御计划（很快被媒体以当时的科幻电影《星球大战》命名），才算有一个大国开始认真考虑将战略武器放到太空。二十年后，乔治·W. 布什总统宣布退出1972年美苏签署的《反弹道导弹条约》，算是进一步推进了这一失败的提议。但是，基于太空的反导弹系统到此时暴露出的低劣性价比已经是非常明显。

68

尽管武器在太空行动中发挥不了多么显著的作用，近地轨道却成为军事行动的关键地带。自冷战时期一直到21世纪，侦察卫星、全球定位系统（GPS）等非武器技术的重要性一直在增加。到21世纪初伊拉克和阿富汗战争爆发时，美国在陆地和海上的行动所需的通信、情报、导航和气象监控已经如此依赖卫星，使得他们意识到：卫星的薄弱保护会对美国的安全构成威胁。因此，在世纪之交，进军太空的这些国家都在开发新技术来保护他们的太空资产，同时也威慑那些潜在的敌人。太空战有一天也许会爆发，但在冯·布劳恩范式用到的技术被取而代之之前不太可能。

近现代作战

美国幽默作家威尔·罗杰斯曾说过，“你不能说文明停滞不前……因为每一场战争他们都会有新的方法来杀死你”。罗杰斯在两次世界大战间隔期提出的这番洞察之语，即便到了21世纪，依然不失其深刻。然而，正如本书所着力强调的，从长期人类历史来看，这话不算准确。在长达数千年的时间里，作战的方式和器械其实演进得极其缓慢。直到中世纪，西方国家发生了火药革命，将化学能加入作战，才开启了一个加速变革的时代。现代作战加快了这一进程，并将其拓展到了四个物理领域：陆地、海洋、天空和太空。

69

历史学家习惯把近现代分为两个阶段。“近代”通常指的是中世纪结束到法国大革命，大概是1500年至1789年。“现代”是自那以后的时期。当然，也有些历史学家把20世纪后半期以来细分为“后现代”时期，不过“后现代作战”这个概念并没有多少解释力，所以我们还是把“近现代”作为一个时期的概念来说明“近现代作战”。

人类学家和社会学家归纳了现代社会所表现出的一些“现代性”特征，其中最为重要的是：启蒙主义的理性、世俗化、大行其道的民族国家、工业资本主义、科技进步，以及针对军事和非军事目标的杀伤力和破坏性强大的作战方式。特别是在20世纪，世界大战的惨烈和核战争摧毁世界的可能性，给这种"现代性"蒙上了空虚和不祥的阴影。人们担心，有朝一日人类会毁灭于自己一手造成的惊天灾变。

但是在19世纪，现代性的好处似乎远远大于其害处，特别是在那些取得飞速"进步"的西方国家。这些进步的核心基本上是物质的，是对物理世界的理解和掌握不断加深而带来的财富、舒适、安全和健康——至少对于创造和定义了现代性的西方国家来说是如此。

科学哲学家阿尔弗雷德·诺思·怀特海说过，"19世纪最伟大的发明就是发明了发明的方法"。他的意思是，西方科学革命之后继之以战争与技术的现象，促进了运用科学方法进行技术发明的做法。问题被分解为各个组成部分，然后对每个部分进行研究：利用现有文献，进行观察、假设、实验、创新和生产。拿到拼图的各个部分之后将其组装成一个系统的子系统，然后再重复同样的过程将其组装到更大的系统中。莱特兄弟发明飞机就是一个经典的例子。当然，这种创新的系统，对促进军用技术和民用技术的发展一样有效。不过，自1815年到1914年，不列颠治下的和平时期加上没有大国参与的战争，让那些正在工业化的国家意识不到作战工业化后会带来怎样巨大的杀伤力。被殖民地区以及少数有先见之明者看到了将会发生什么，但大多数西方人还只是把作战的现代化视为他们文明先进的又一项指标。

我们之前讨论过的海战从帆船时代到蒸汽轮船时代的转换，可以作为19世纪技术创新的案例研究。从最初的蒸汽船（罗伯特·富尔顿1807年造的"老北河"号）到19世纪90年代英国的"威严级"战列舰，军舰的火炮、船体、装甲、推进器和军舰的大小所发生的变化，都需要工业、金融、政府等各种基础设施的支持，而这只有最发达的国家负担得起。即使强大如法国，最后也在20世纪初退出了海上军备竞赛，只留下英国、德国、日本和美国这几个真正的海权竞争者。雄心勃勃的俄国一度认为，他们落伍的舰队尚可一争高下，结果在1905年对马海峡的海战中遭遇了史上最大的败绩。辅助性技术的出现，比如鱼雷、无线电、陀螺仪、高压涡轮等，进一步增强了海洋强国的战争能力。

火药驱动的武器——这一最具革命性的军事技术，也在发生巨大的变化。无论在陆地还是海上，原先从外部点燃、内膛光滑、单次击发的火炮和轻武器，都让位给了自驱动、连发、有膛线的枪炮。有膛线的火炮比滑膛炮射程更远，也更准，有自击发装置的预包装炮弹（或火药袋）可以用电或撞针引爆。在手枪当中，可以连续发射而无须中途装弹的左轮手枪，到1900年被装了弹匣的手枪所取代。步枪也同样采用了弹匣，要么通过螺栓和杠杆原理用人力装填，要么利用气体膨胀产生的后坐力原理自动装填。完全自动击发的机关枪在19世纪末出现，从美国南北战争时期准现代的手摇式加特林机枪，发展成1899—1902年的波尔战争中大显神威的马克沁机枪。擅长发明省力技术的美国人，率先大规模地生产这些致命武器。

19世纪陆战发生的技术变革，与中世纪末期引入火药产生的变革大不相同。在这两个时期，火力都是制胜的关键，但增强火力有不同的方式。中世纪末期和近代早期的枪炮粗大笨重，对敌时发射和装填都很缓慢。每次发射完毕都需要清理枪炮膛管，接着再装入火药、填料和密集弹丸，然后装上引线或起爆药，再从外部点燃——遇到风雨，有时还发射不了。

这样的装备要提高火力，只能在训练枪炮手上下功夫了。给轻武器装弹，在16世纪需要九十多个步骤。一队步兵完成一次齐射，其速度取决于动作最慢的那个枪手。而敌人的骑兵，这时可以等在火枪射程之外，然后趁枪手装弹间隙纵马冲杀。因此，16世纪的步兵阵型中，往往在枪手前面安排长矛手以保护装弹的枪手；而那些训练有素、动作整齐的枪手，就能以最高的频次发射火力。

到了19世纪，火力的发射频次来自枪炮本身而不是枪炮手。现代技术的两个特征——机械化和自动化——合在了一起，让战场上弹如雨下。随着技术的演进，对士兵的技术要求有所下降：他们只需瞄准和射击，射得多比射得准更为重要。这种枪林弹雨的“阿喀琉斯之踵”在于供应，即后勤能否为枪炮提供足够多的弹药。

19世纪，欧洲的化学家发明了一种可用于枪炮的推进剂——美国人称之为“无烟火药”。之前使用的是黑火药，燃烧不充分，所留下的固态残留物容易堵塞膛管，产生的浓烟能笼罩陆上的战场和海上的甲板。19世纪新出现的这种推进剂大多基于硝化纤维（火药棉），与其他配方结合，可以大大提高火药的稳定性和爆炸力。新的推进剂减少了烟雾和残留，增强了枪炮的火力、射程和可靠性。装弹标准化后，可以计算出更准确的射表。火炮因此可以快速布局，实施“有效射击”。不难想见，火炮的这些改进会促使它向“巨型化”发展，作为攻城武器和舰载火炮都是如此。实际上，19世纪末出现的现代战舰就发现自己卷入了大炮和装甲的“矛盾之争”，这种技术上的相互克制将一直延续到第二次世界大战。

非武器的军事技术也为19世纪的作战转型贡献了力量。蒸汽轮船——我们应当还记得它最初是一种非军事的商用技术，在这个世纪里也慢慢找到了在海军舰船上的用途。自1807年罗伯特·富尔顿的处女航，到1862年“莫尼特”号和“梅里马克”号在汉普顿锚地第一次发生汽轮舰船对决，时间上间隔了半个多世纪。铁路这一民用发明，也是同样的模式。它最初的用途是为了将煤从煤矿运输出去。到美国内战时期，它们被用来在战场间转运军队和补给。战前铺设的铁路网利于北方：北方的线路横跨东西，而南方连接中心和外围的线路设计恰恰为北方军队的进攻提供了通道。在德国统一战争期间（1864—1871），出于军事战略目的而设计的铁路网大大便利了普鲁士军队，既能把部队和装备运到前线，也能把伤兵运回家。

19世纪的通信技术也被用作军事和民用场合。电报让掌控战场全域行动的指挥能力得以扩展，能让远离战场的军职或文职领导人实现远程调兵遣将。19世纪末期铺设的海底电缆，让指挥官的命令可以跨越国界，从而避免了类似1815年新奥尔良战役中的悲剧再度发生（当时结束1812年战争的协议已经签署，但还有大约336名英国和美国的军人在此后丧生）。

新技术层出不穷的19世纪也涌现出无数非军事的技术创新。拿破仑的军队曾尝试使用“可更换零件”——这是美国发明家伊莱·惠特尼在19世纪头十年里改进的技术。尽管惠特尼的零件还需要手工打磨之后才能实现真正的互换，但他的榜样力量激发后来者不断地改善技艺。三千年来，人类一直试图制造钢材，但直到有了一系列冶金的发现和生产的创新，才使得

74 机械化、工业化地生产大量各种形状的定制钢材成为可能。19世纪末期的工业巨头，如克虏伯、卡耐基等人，通过生产火车头、摩天大楼、军舰、大炮等需要的大量钢材赚取了巨额财富。事实上，能够为工业基础设施生产大量钢材已成为经济和军事实力的标志。另外一个不起眼但同样具有变革意义的发明，是日常见到的锡制罐头——不光方便地给世界各地的人带来了各种食物，也为行军中的士兵们提供了营养。法国的蒙戈尔费埃兄弟发明了载人热气球，很快它就发挥了作为侦察平台的军事用途。有了这样的基础，飞机的发明还会远吗？

这些快速发展的技术给作战带来的变化，对帝国的殖民战争产生了最大的影响，因为它赋予了西方列强对土著居民的非对称优势。当然，自15世纪肇始的第一波西方帝国主义浪潮算起，西方国家就已经享有了这样的优势。比如，西班牙探险家埃尔南·科尔特斯，就曾用舷炮帆船把西方的力量投射到墨西哥海岸，随后征服了整个阿兹特克帝国。随他夺取阿兹特克首都的仅仅是一支数百人的部队，但是装备了火药枪、马匹以及在他们当地造的炮艇。实际上，当地盟友对科尔特斯的帮助，与军事技术给予他的帮助一样大。然而，阿兹特克人并不是最后一批被装备了火器的军队所震慑的土著人。

历史学家丹尼尔·海德里克曾指出，第一波的帝国主义扩张使得欧洲列强在1800年现代时期开始以前，就控制了全世界大陆面积的35%。而在漫长的19世纪中，同样是这些西方列强，通过运用新的技术，使他们的控制面积又增加到了84%。在一75 个世纪的时间里征服世界上一半的大陆，按照海德里克的说法，其关键就在于向内陆投送力量。16世纪科尔特斯对墨西哥的征服，算是一个例外。近代欧洲人的对外征服大多依靠舷炮帆船：它可以驶入那些主要贸易国的港口，控制这些非工业化国家的进出口通道。凭借枪炮的威力，他们可以任命总督，确保当地的统治者按照这些殖民国家的商业利益开埠通商。这些殖民者并不需要占领被殖民的国家，就可以实现他们的资本主义目的。

海德里克认为，技术的变化在19世纪改变了这种模式。蒸汽轮船，让西方殖民者可以将海军力量沿着江河投送到内陆。电报，让身处港口或首都的总督们能够与内陆前哨保持联系。火力强大的新式枪炮，使得西方的小股部队也能够凌驾于当地使用古董冷兵器的大部队。铁路，可以进一步将西方军队运送到不通江河的贸易中转点。苏伊士运河，缩减了欧洲各国与其在非洲和东南亚殖民地的路途。海底电缆，让海外总督能与本国政府保持联系。奎宁，虽然称不上是技术，但也让殖民者免于染上当地一些让人极度衰弱的疾病。总之，19世纪的技术，使得欧洲的帝国主义者能够实现对殖民地区域和人口的整体掌控。

技术在19世纪殖民战争中所发挥的决定性作用，在列强之间的战争中却是大打折扣。部分原因在于，这些强国——主要是欧洲各国加上美国以及19世纪晚期的日本——在技术变革上大体同步。工业强国的陆军和海军装备处于对称性发展，它们 76 一般体验不到殖民战争中的压倒性优势。更为重要的是，19世纪的列强之间并未爆发真正的大战。自拿破仑1818年在滑铁卢战败，一直到1914年第一次世界大战爆发，世界一直处于“不列颠治下的和平”。这反映出英国对于各大洋的牢牢控制，以及欧洲在长达四分之一世纪的法国大革命和拿破仑战争后的精疲力竭。但这种模式下有两个例外，也预示着在工业化国家之间发生战争时技术对于作战方式的改变——尽管没有多少人意识到这种改变会有多么彻底。

1861年至1865年的美国内战，创造了很多个“第一”。北方的工业优势大大超过以农业为主的南方：他们有先进的运输和通信网络，还有批量生产的基础设施，这些很快就转化成战争产能；而南方邦联这时还只是从零起步。北方有一支海军以及为其提供支持的配套和岸边设施；南方采取了一些创新措施与之对抗，比如：封锁水路，袭击商船，布设水雷，实验鱼雷和潜艇，等等，但最终却未成气候。北方在应对之余，也借鉴了南方的这些创新，并把装甲炮艇加入他们的武器库里。战争的第一年，双方都以极大的热情把蒸汽机驱动的装甲军舰（分别为“莫尼特”号和“梅里马克”号）投入实战。这场发生在“汉普顿锚地”的战斗，被视为海军由帆船时代转为蒸汽机时代的转折点。北方在人口和财富上具有对南方的压倒性优势，这些优势又被领先的技术加以放大。技术为其提供了今天所谓的“力量倍增器”。

1864年至1871年的德国统一战争，则为变革军事技术提供了另一个试验场。在与丹麦、奥地利、法国进行的一连串战事中，普鲁士军队的迅捷果断和所向披靡让世人震惊。同大多数重大历史事件一样，普鲁士的成功背后也有很多原因，不仅仅是由于普鲁士的军国化，敌人疏于准备，普鲁士军队训练有素，或者普鲁士的宰相奥托·冯·俾斯麦（1815—1898）在地缘政治中施展的手腕——他成功地孤立了对手，并防止了外来干涉。技术，在其中也发挥了非常实际的作用，特别是发挥了战略用途的铁路，以及高性能的普鲁士德莱塞击针枪。这种枪让士兵能以卧倒的姿势装填子弹。但到了普法战争时期（1870—1871），

77

普鲁士军队的这一项发明就优势不再——法国人发明的后膛步枪被证明与德莱塞击针枪一样高效。而这也正是工业化条件下现代作战所表现出来的一个特征，即技术相互促进、相互制约。轻武器上的这种势均力敌，某种程度上抵消了一方希望通过技术创新所获得的独家优势。

19世纪中期发生在美国和欧洲强国的这两场战争所展示的技术力量让人惊叹，也符合先进的人类可以征服大自然和落后地区这种普遍的观点。这个世纪的最后几十年里，西方人大肆操办了十几场国际性展会，以此彰显他们在技术上的超凡。1851年的伦敦世界博览会首开先河，各国随后纷纷效仿。用历史学家迈克尔·阿达斯的话来说就是，欧洲人将机器视为“衡量人类的标尺”，它证明了他们文明的优越性，以及按照他们的标准征服和改造世界上其他地区的正当性。

到19世纪晚期，现代武器不断增长的致命杀伤力引起了人们的重视，由此引发了对军备控制的兴趣。这个话题贯穿了整个20世纪，并延续至21世纪。有史以来，甚至在更早的时期，人类社会就达成一种共识，即，如果战争对象被视为同类时，某些军事技术和手段的使用应加以克制，但如果战争对象是陌生人或野蛮人或“另类”，则不受此限制，比如著名的“十字弓限制令”。1139年天主教的拉特兰会议就规定：十字弓不可用于基督教徒，但可以用来对付穆斯林。1899年和1907年的《海牙公约》就对下列武器和行为做出限制：使用毒气，使用入身变形子弹，从气球上抛掷投射物和爆炸物，武装商船，以及敷设自动触发水雷，等等。实际上，这类限制会因为“军事必要性”被弃之一旁，令其有效性大打折扣。只有符合自身利益，参战各方才会遵守，这时军备限制才能真正发挥作用。

在19和20世纪之交，只有那些最为睿智的时代观察者，才能洞悉军事技术的演进将把人类带向哪里。其中最为杰出的是波兰银行家和金融家扬·布洛赫（1836—1902）。他在研究19世纪作战的多卷著作中就预言：战争注定不再会速战速决；在轻武器和大炮不断增强的火力下，士兵们只能伏于地面作战，战场机动变得不可能；军队会增大规模、增强火力；战场会超出指挥官的视野和理解范围；工业会给庞大的军队提供无限的食物和弹药供给。布洛赫认为的最终结果就是：战争变成了静态的耗损战，谈不上以胜利而告终，只能是战争双方在道义和经济上相互耗损。而这个僵局的核心，正是技术。

总体战

布洛赫的不祥预言在第一次世界大战（1914—1918）中不幸成真。西线堑壕战中，战场上弹如雨下，士兵们只能藏身于蜿蜒曲折的战壕，战壕从瑞士一直延伸到大海。战争双方都想方设法打破僵局，尝试运用各种战略、战术、技巧和技术，包括：火炮齐射，使用化武，攻击商船，战略轰炸，登陆作战，运用火攻，以及使用坦克。但这些都没有用。正如布洛赫所预言的，双方在道义和经济上的损耗才最终决定战争的结果。

79

1939年至1945年的第二次世界大战，有时候也被视为发生在大国之间的同一场世界大战的第二阶段，只是规模更为庞大，对世界的改变也更为深远，堪称人类历史的一个分水岭。第一，这两次冲突是人类仅有的总体战；第二，战争比拼的是工业产能，最后的赢家是能够生产最多产品的国家联盟；第三，第二次世界大战是第一次在四个作战领域——陆地、海洋、天空和太空——全面展开的战争；第四，第二次世界大战末期与战争之初所使用的武器有显著的差别——这是人类历史上的第一次；第五，第二次世界大战是最后一场大国间的战争；还有第六点，第二次世界大战以核武器革命而结束，这在战争技术和人类历史上都是一个转折点。

记者兼历史学家沃尔特·米里斯（1899—1968）认为，“总体战”可谓三次伟大的历史性革命之集大成者。法国大革命，首创“全面动员”的概念，全民得到武装；战争与技术，展示了生产战争物资的能力，可以为庞大的军队提供武器、装备和运输；福特的生产线及20世纪的大规模生产，又在工业革命的基础上提升了速度和效率；而普鲁士总参谋部首创的组织管理革命，可以实现部队的跨时空集结调度。只有在上述这些能力都得到实现之时，演进了五千年之久的军事技术，才在世界大战中迎来了屠杀和破坏的顶峰。

这两场世界大战中比拼的是交战国的工业产出，它集合了工业现代化和20世纪的大规模生产能力。在战争过程中，每一方都期望凭借自身的强大军火库，不仅要摧毁敌方的军火库，还要摧毁敌方的意志和物质资源。因此，敌方的民众无可逃避地成为攻击目标，因为他们既体现了国家意志又表现为生产力。历史上从未有如此多的物质被用于战争，也从未有如此巨大的破坏。在人类历史上，战争中的死亡多由疾病、饥饿和流离失所造成，而并非死于直接的攻击；世界大战则与此不同。屠杀和破坏的能力已经被工业化了，交战国的损毁程度前所未见。第二次世界大战中的轴心国，最终在耗尽民众意志之前，就耗光了物资。

80

人类历史上的战斗大多发生在陆地上，直到古代末期或者说古典时期早期，国家间的冲突才发展到海上。又经过了两千年的技术进步，人类才第一次飞了起来，从第一次飞行到第一次空战又过去了十多年。但仅仅在几十年之后，太空战便接踵而来：V-2火箭后来的确飞入了太空，尽管在第二次世界大战期间它的飞行轨迹还仅仅局限于大气层之内。在不到半个世纪的时间里，两个新的作战领域就被开辟出来，这愈发加重了人们对技术失控的担忧。一些现代军事技术研究者把网络战归于第五类作战领域，但其中的电磁控制原理并不新鲜，早在第二次世界大战时期就出现了。

第二次世界大战见证了很多1939年前未曾出现的重大军事技术的研发和使用，其中包括：微波雷达、喷气式发动机、近炸引信、制导飞弹、巡航导弹、“精确”投弹瞄准仪、音响鱼雷、计算机破译密码，当然还有——原子弹。这里值得关注的重点是，第二次世界大战中出现了系统化、制度化的军事技术研发。

第二次世界大战是最后一场大国间的战争。实际上，自1945年以来很少发生国家间的战争，战争大多都是国内战争：叛乱、暴动、内战和失败国家出现的无政府状态。这些冲突很少会像总体战那样，动用整个国家的资源。作战运用的技术一般是传统意义上的，视具体情况而定。换句话说，采用的还是第二次世界大战中的“联合作战范式”：步兵、炮兵和“骑兵作战”（坦克、运兵车及后来的直升机），再加上战术火箭和空中支援。作战技术上的这种停滞，可以归因于1945年来大国间战争的缺席，以及最具革命性的军事技术——核武器的出现。核武器在第二次世界大战即将结束时出现，开启了“长久的和平”（约

81

翰·刘易斯·加迪斯对大国间无战事现象的概括语)，并延续到21世纪的第二个十年。很多因素导致了这个“长久的和平”：传统战争在工业化后产生的巨大破坏性，联合国这样的国际组织的出现，民众对法治日益增长的信念，愈加紧密的国际联系，加速增长的通信和交通，以及越来越多的人意识到现代战争再也没有赢家。

但所有这些因素，都没有被炸后的日本广岛和长崎来得清晰直观。因此，尽管在第二次世界大战之后的二十年里，超级大国从原子弹发展到了热核炸弹（更轻的原子聚合反应，成本更低但威力更大)，核技术也扩散到了其他国家，人类社会却形成了一种对核战的禁忌。人类终于开发出了一款因为过于恐怖而无法使用的武器。如果说在第二次世界大战之前，人们对赢得现代战争还心存幻想的话，那么广岛和长崎则赤裸裸地展示了它的后果。因此，即便两个超级大国的核武库里积攒了七千多枚核弹头，“永远不要使用它们”多少还算是一种共识。

核能时代到来有七十年了，这种共识依然存在。核武器没有消失，并且还在缓慢地扩散。但到目前为止，它们主要还是国际和平的保证。过去的七十年，没有爆发大国间的战争，只有零星的国际冲突，而且是在国际合作的框架之内。当然，这种长久的和平随时可能终结，对核武器的禁忌也会失效。也许有一天，自杀式恐怖主义的信奉者会弄到一件原子或热核武器，或者其他的某种大规模杀伤性武器，但是一旦在怒火中引爆这样的武器，他们无疑会发现：核武器能用来威慑和报复，但用来发动进攻其实并无作用。至少就此刻而言，核武器革命带来了更为和平的时代，尽管在每天的晚间新闻里依然还能看到种种暴行。

82

83

第四章

技术变革

研　发

第二次世界大战不仅产生了新的武器，还带来了世界与军事技术关系上的两个重大改变。一个是之前提到的核技术革命——后面我们还会再度讨论；另一个军事技术的巨大改变，是形成了现代化的、制度化的、常规化的研发。

第一次世界大战中也动员了科研力量，但它主要比拼的是工业产能。第二次世界大战在很多方面延续了这一模式，美国就发挥了填充民主国家军火库的作用。美国的GDP超过了它的主要盟友（英国、法国和苏联）和联合国其他成员国的总和。到战争结束时，这些国家的GDP总和达到了轴心国GDP总和的五倍。北大西洋的战事，到1943年上半年变得有利于同盟国，其原因就在于：同盟国制造船只和货物的速度超过了被德国潜艇击沉的速度。1944—1945年的冬天，德国在关键的阿登战役中失利，由于燃料缺乏不得不将坦克丢弃在战场；而盟军早已铺就了一条穿过英吉利海峡的输油管道，可以为庞大的坦克和卡车部队进军柏林提供足够的燃料。拿破仑的军队行军只能靠体力支撑，20世纪中期的军队可以乘坐舰船、飞机和车辆，而这些靠的是内燃机和燃油支撑。所以他们身后的后勤补给线很长，如脐带般一直延伸到美国总统罗斯福的军火库。 84

同盟国——尤其是美国——的工业能力赋予这个多头的庞大机构巨大的能量，但产出的战争物资却未必总是最好的。比如，美国海军使用的飞机和鱼雷，比日本产的要差；美国的坦克劣于德国和苏联的坦克；德国和英国的喷气式飞机在战前就飞上了天，而美国没做到；德国的长航程潜艇与美国的比毫不逊色。到了战争的后期，在同盟国的重压下，德国陷于崩溃，希特勒把他日渐萎缩的资源投入到秘密武器的研究，新研发的技术险些改变战争形势；他的喷气式飞机——尤其是ME-262型战机——拥有抹去盟军空中优势的潜力，但因为还存有不少缺陷，而且燃料供应不足、飞机数量缺乏，所以终究未能对盟军造成真正的威胁。韦恩赫尔·冯·布劳恩研制的火箭可以打到英国本土，但准头不行且数量不够，并不足以让英国屈服。然而，这些新武器的潜力足以表明，同盟国在军事技术的研发创新上并不处于垄断地位。在广岛和长崎爆炸的原子弹，最终结束了太平洋战争，也证实了科技研发对于战争的决定性影响。

军事技术多次落于人后的经历让美国决意改变。虽然战争期间他们在科技上不乏显著的成就，比如杰出的“曼哈顿项目”，但军方领导人开始认识到，不能再沿用战前的机制来研发新的军事技术了。数量曾经是大战中获胜的主要因素，但下一场战争的结果将会由质量所决定。 85

美国负责战时科技力量动员的是万尼瓦尔·布什，他是战时科研办的主任，也是富兰克林·罗斯福总统实际上的科学顾问。第二次世界大战结束时，他为总统写了本书——《科学：无尽的边疆》（1945），书中为美国政府支持军事、医药和经济等领域的研究创新设计了蓝图。布什在第二次世界大战中的经历让他确信，科学家才是真正懂行的人。政府应当出资成立国立研究机构，并让科学家来主导设立科研项目。美国政府自然不愿意放任一个组织自由发挥，所以拒绝了他的提议，代之以成立国家科学基金会（NSF）和国家卫生研究院（NIH）来分别进行科学和医学的基础研究。其他由政府资助的科研大多由政府各部门负责，比如新成立的国防部。国家科学基金会和国家卫生研究院从事的是基础研究，追求的是推动整体科技水平的提高，而各政府部门则从需求出发，把科技成果转化成具体的用途。

在国防部内部，三大军种——陆军、海军和刚刚独立成军的空军——很快就发展出了适合它们各自目标和理念的独特创新机制。三军中技术最弱的陆军，依然保留战时的合作方，比如致力于计算机开发的宾夕法尼亚大学摩尔学院。陆军还为总揽研发活动进行了内部设置，最初是一个研发室，后来是研发主管办公室。除此之外，它还是继续依靠历史悠久的军械体系来进行创新。

海军被证明是三者中最具进步性的一支，它充分利用了长久以来与从事基础研究的科研院所建立的联系。它延续并扩大了战时的海军研究室，强化了位于华盛顿特区的著名的海军研究实验室。这些机制补充了海军已有的军舰设计开发项目，最终演化成了海军海洋系统司令部。

空军源自战时的陆军航空兵部队，采用最引人注目的举措形成了一种新的政府支持下的创新模式。在延续战时依赖研发合同方式的同时，他们招募了加州理工学院传奇的空气动力学专家西奥多·冯·卡曼，请他主持一个咨询委员会，并做出了多达十二卷的关于未来航空的研究。第一卷由冯·卡曼亲自拟定的题目——《科学：称霸天空的关键》，就说明了一切。空军继续以合约方式购买发明，甚至还资助了兰德公司——美国军事研发后来所倚仗的智囊团中的第一家。空军沿袭了陆军的军械所传统，同时扩展了在俄亥俄州的莱特试验场的内部研发，并开设新的实验室，比如，位于田纳西州塔拉霍马的阿诺德工程开发中心。

各军种研发新技术的热情如此高涨，让国防部长觉得有必要设立管理机构来限制种种创新乱象。1947年的《国家安全法案》不仅创设了空军这一军种和设立了国防部长的职位，还成立了研发委员会。该委员会在艾森豪威尔政府变成了助理国防部长的职位，虽然用过不同的名字，但自1953年起就这样确定下来了。苏联人成功发射的第一颗人造地球卫星，引发了美国的强烈反响，各军种提出了五花八门的太空计划。艾森豪威尔总统因此认为，有必要再设立一个机构，来梳理各种为占得技术先机而相互竞争的方案。1958年，国防部的高级研究计划局就此成立，负责审核各军种提出的尚未完善的研究计划，同时还发挥另一项作用，即寻找并开发在他们看来符合国家利益的新技术。

在各个机构的大力推动下，冷战期间的国防研发一哄而上，产生了一些技术过头、大而无当的项目。苏联的人造地球卫星上天之后，美国陆军提出了在月球建立基地的建议——因为军事理论教科书总是主张“占据高地”。20世纪50年代，陆军和空军竞相开发火箭技术，即所谓的“托尔-朱庇特之争”，耗费巨资研究各自版本却大体相同的中程弹道导弹。各军种之间围绕国防资金的争夺激发了技术创新，因为创新能带来机构设置和预算拨付上的优势。海军陆战队坚持开发的能垂直起降的飞机——“鹞”式和“鱼鹰”攻击机，也被证明造价昂贵却实用性不强。海军孜孜追求的是核动力舰队，但对除了潜艇和航母外的大多数舰船来说，这种动力系统耗资太过巨大。空军试图开发一种可以反复使用的载人飞船，他们将它命名为“Dynasoar”，取的是“一跃冲天”的意思，但这个词的发音与“恐龙”（dinosaur）相同，结果倒是不幸应了这个物种的灭绝命运。空军21世纪重金打造的F-35多用途战机，也正在耗尽他们的经费。

1961年，艾森豪威尔总统在他的离任告别演说中，将这种热衷研发的气氛贴上了“军工联合体”的标签。他的意思是，通过夸大冷战威胁，国防合同承包商和军方围绕共同利益形成了紧密联系，允诺通过昂贵但尖端的技术来实现安全。后来有很多观察者提出，美国国会和高校也是这个“军工联合体”的同谋。国会议员发现在自己的选区开展军事研发和生产有利可图，高校也乐于接受军方的研究经费。美国此举也非独此一家。一位历史学家就称冷战时期的英国是一个“战时国家”，而加州大学伯克利分校的一位政治学者认为：冷战正酣时，美国也许存在“军工联合体”，而苏联实际上就是一个“军工联合体”。

总之，大国为了获得新的更具优势的军事技术，竞相制度化军事创新。这也导致一种情况：在新的武器系统刚刚投入使用时，下一代的技术已经开始研发了。这种刻意的更新换代，效仿的是20世纪50和60年代美国汽车界每年发布新车型的做法。“性能贪欲”成为对武器系统的主流需求，导致武器的生产成本上升和可靠性下降。这些工业国的军事机构发现，与其说他们在与其他国家的对手竞争，不如说他们更多是在和自己竞争。研发动力一部分来自国际军火市场——那里对最新的军事技术总有无尽的胃口；但更多的动力，则来自国内不断强化的军工联合体。

时间的流逝解决了一些分歧。冷战结束了，终究没有引发众人恐惧的第三次世界大战，世界又回到了约翰·路易斯·加迪斯所称的“长久的和平”。但无论是在战争还是和平时期，军方在美国政府用于研发的开支中都占了大头。这种现象在很多方面存在争议。耗费巨资去开发问题多多的军事技术挤占了基础研究经费，而后者能够产生长期性、根本性的创新。很多经济学家认为，军事研发对经济增长的推动作用不如直接将经费投资于民用领域，如能源、交通和基础建设。民用研发更可能产生军事用途，而不是反过来。而且，军事研发存在更大的牟利空间，因为合同双方所处的市场里面买家单一、卖家寥寥。唯一的买家倾向于“性能贪欲”，而几个卖家也不会相互压价。

89

军民两用技术

除了既可军用又可民用之外，一些军民两用技术还有一重意思：军方既把它用作武器，同时又发挥它非武器的作用。本书提到了一些例证，比如，城防工事、道路、二轮战车、蒸汽机、运输机、核能等。这类军事技术值得仔细研究，因为广义技术背景下的作战技术，为我们打开了探究军界和民间辩证关系的一个维度。正如民间社会理所应当获得军队的保护，他们同时也理所应当地获得了军事技术。不仅如此，很多的民用技术脱胎于军方，后来深刻地影响了民间社会——这一点很少有人研究。

非武器性军民两用技术

非武器性军事技术，显然最适合成为军民两用技术的候选者。它们为作战提供支持，不直接攻击人或物。作战中一个无法阻挡的趋势是，非武器性军事技术在数量和重要性上不断增强。最早的作战无疑开始于最简单的器具——矛、刀、棒、石头和弓箭，全都是武器，也不怎么需要支持。但慢慢地，人们发现，拥有盔甲、后勤、情报、通信、医疗、交通等支持的战士更容易打赢。

90

随着这些服务保障不断拓展，战士们渐渐地被当成“长矛之尖”，而长矛杆的重要性不久就超过了尖。到了21世纪，为作战提供支持的人员和物资能占到军队的九成。用现代术语来讲，这就是“牙齿与尾巴的比率”——前线士兵与支援力量之比。尽管军事文化还继续凸显战士们冲锋陷阵、打击目标的重要性，事实却是，非武器性技术在数量和重要性上都超过了“长矛之尖”。我们可以通过几个例子来说明这一点。

城防工事——这个最具影响力的非武器技术，我们前面讲过。它主要不是决定战争中谁赢谁输——尽管有时候也有这个效果，它更多是决定一些战争什么时候会发生，或者更重要的，避免战争发生。为了同依然生活在荒郊野外的野蛮人和游牧者区隔开，一些国家和文明建造了城市，在那里耸立着包括城墙这样的高大建筑。城墙当然只是人造物品，不是技术。但是，它与那些寺庙、神塔和会堂等高大建筑分享同样的建造技术。无论是用石头（杰里科古城）还是干燥砖（乌鲁克古城）或混凝土（罗马古城），这些城市打造的城墙总能将野蛮人拒之门外。从设计之初，城墙就是为了阻遏现实和潜在的进攻者，例如，保护君士坦丁堡长达一千一百多年的城墙。它宣告的是，城内的居民拥有足够的能力和资源来抗衡任何外来挑战。总之，它是对战争的抑制，是对来犯者注定徒劳无功和失败而归的承诺。各个文明在扩大军事力量相互征伐的同时，也在不断加固自己的城墙，他们对其他国家发出的正是这样一个信号。

有些国家加固的不仅是自己的城市，还有边境线的薄弱地带。中国的长城，耸立了一千多年，沿着中国北方边境曲折蜿蜒一万三千多英里。罗马人也在天然边界和入侵路线上建造了他们的边境墙。最初是在道路上间隔设立防守瞭望塔和堡垒，多用栅栏，偶尔也有土石筑造的墙，如英国的哈德良长城。同中国人一样，罗马人并不指望单靠城墙挡住敌人，而是迫使敌人改变进攻方向和拖住敌人的进攻步伐，以便调集自己的军队到边境抗击。第一次世界大战后，法国人建造“马其诺防线”的初衷与这种想法其实没什么不同。虽然“马其诺防线”在1940年被入侵的德国军队轻易绕过，因此给静态防御带来了不好的名声，但它实际上还是实现了当初的意图：减缓敌人的入侵速度和使其改变方向，直到己方增援力量到来。不过，对法国人来说，不幸的地方在于，他们的军队没能守住防线。

历史上的城防工事也给了建造者们一项额外的收益——拥有城墙的国家借此可以减少用于防守的常备军队。简单来说，作为和平时期的安全投资，城墙在建成之后的年份里依然享有安全红利。那些能够能让国民出钱出力建造公共防御工程的国家，可以省下一大笔和平时期维持常备军的昂贵军费。同时，由于城防工事没有什么进攻性，是一种对和平的投资，所以算是一种并不直接威胁到邻居的军事技术。

道路，就其历史性和重要性而言，大概是排在第二位的非武器军事技术。同城防工事一样，道路不是技术，而是技术制品。事实上，最初的道路甚至算不上是“路”，只是一些通道，比如“丝绸之路”，是人和动物频繁往返留下的痕迹。但不久之后，一些文明国家开始改善这些自然形成的通道。当这种改善达到了使用工具和器械，运用一定工艺做出坚固耐久的路基的阶段，道路的技术就产生了。考古发现，这样的道路早在古代波斯、中国、秘鲁以及其他帝国就出现了。罗马人把这项技术提升到艺术的高度，无论是建造古斗兽场、高架渠，还是纯军事用途的野战工事和桥梁，都展现了同样高超的工程能力。借助路面硬化的道路，他们将帝国各地缝合在一起。其中一些修筑精良的道路一直存留到21世纪，跨越了近两千年。这些道路与其现代的变种，如德国的高速公路系统和美国的州际高速公路系统相比，其共性在于：他们服务于商贸、政府管理等非军事用途，但同时也便于国家动员和调动军队去抵御外敌。但不幸的是，历史上许多国家的道路也为入侵者提供了坦途，结果反而与其当初的军事目的背道而驰。

更近的一个非武器性军民两用技术的例子是蒸汽机。蒸汽机的发展是技术推进的一个经典案例。它源自自古以来的科学好奇心，直到18世纪出现了这种为煤矿矿井抽水的商用工具。然而，这些早期的蒸汽机效率不高，而且也不经济，只能用于煤矿的坑口——因为那里的燃料便宜。直到1769年詹姆斯·瓦特

92

发明了分离式冷凝器，蒸汽机才开始真正发挥潜力。布尔顿-瓦特公司同商业伙伴、火炮制造商约翰·维尔金森合作，大大地改良了蒸汽机：布尔顿-瓦特公司为维尔金森钢铁厂的钻孔机提供动力，而维尔金森为布尔顿-瓦特公司提供精密镗床加工技术，这才造出了蒸汽机的密合缸体。民用和军事技术的这种密切协同，在任何军工联合体中都很罕见。很快，蒸汽机就不仅仅为工业革命中的英国工厂提供动力，还被用于铁路。火车在美国的南北战争和德国的统一战争中被用来运送部队，装了蒸汽机的军舰也有了足够的动力战胜风力和潮汐。即便是最现代的使用燃油或核动力的主力战舰，其实也还是蒸汽机船，靠的是新式汽轮机里面的蒸汽推动自身和其他辅助装备。

93

内燃机的出现，对于作战和民间社会同样重要。第一台内燃机是火炮，是人类掌控密闭空间里碳化合物迅速燃烧产生的能量，用来实现自身目的的器具。蒸汽机是外燃机——它使用外部的火来加热密闭空间里的水；而能将火产生的能量直接转化成机械能的实用机器，直到19世纪晚期才出现。19世纪人们围绕内燃机的一系列实验，在石油馏分技术投入商用以后大大加速，最终生产出了用火花和压缩点火的实用内燃机。

到第一次世界大战时，内燃机已经用于军机、潜艇、卡车、坦克，甚至可以提供辅助电源。只要内燃机和燃料能到的地方，就会有电。发的电可用于照明、供热、无线电、电报、修理厂、医院、厨房、冰箱，以及无数可以用来支持军事行动的电器。自19世纪末期，巨大的固定发电厂就开始为城市提供电力。随着由内燃机供能的移动发电机被开发出来，军队可以在战争中运用各种现代设备了。飞机可以载着乘员所需的无线电等各种设备和辅助设施，投入格斗、轰炸和侦察任务。内燃机为总体战赋予了能量。

94

另一项还在民间社会广泛使用的非武器军民两用技术是电力和电子通信。它涵盖了从最早的电报到更为现代的电话、无线电、电视和互联网。这些新近的通信方式能近乎光速地以模拟和数字信号来传输编码、声音和图像。当然，历史上使用过的烟雾和旗语也能以光速传递信号，但它们都受限于视线范围。（声音也能以自身的速度传播，但比光速要慢很多。）自19世纪开始，军事人员只要连上电话线就可以接近光速地互相交流信息；无线电的出现进一步加快了信息交流，而且它不再需要铺设电话线，尽管它也受限于技术和环境因素。

在现代的数字通信中，所有内容都已经数字化了——被转换成二进制的形式，然后以适当的电磁波传输，最后在接收端被转换成数据、声音或图像。根据接收端的情况，它可以光速在视线范围内进行传输或广播。战争是一场零和游戏，一方的优势就是另一方的劣势。一位指挥官如果在己方情况暴露给对方之前就知晓敌方动态，并且比对手更快地将命令下达到下级，就会在战场上占据压倒性的优势。今天的军事指挥官所拥有的通信能力，可以让各种信息几乎瞬间实现全球范围的上传下达，还可以在战斗中与下属实时交流。不知道这算是好事还是坏事？随之而来的一个有趣问题是：这种能力消除了克劳塞维茨所说的“战争迷雾”，还是让迷雾变得更浓了？

“Computers”（计算者/计算机）这个词，最初指的是第二次世界大战前美国陆军部队里从事弹道曲线计算的那些女性文职人员。到了第二次世界大战时期，计算机——又一项非武器军

民两用技术——才变成机器。这些机器到了21世纪发展到无所 95
不在的程度，不能仅仅归功于军方或民间单方面的推动。无论如何定义“计算机”，这两方面都做出了不可或缺的贡献。在21世纪，人们的生活中发生了计算机革命——或者正在发生，这一点看似司空见惯，但对于哪些是革命性的变化，人们的看法并不一致。它是通信、信息、计算、人工智能抑或仅仅是娱乐方面的革命？可能最好还是在技术语境下来思考这个问题：把计算机看成改进过的固态电子设备，是它使得人类活动中所有这些领域的巨大改变成为可能。

从这个角度看，为了实现计算弹道射表、通信加密解密、模拟核反应以及整合雷达网络等目的，军方对早期的模拟和数字计算机研发做出过重要的贡献。第一支晶体管出自民用电话交换机中，但微处理器的两大发明人之一杰克·基尔比，是在为美国空军开发导弹的电子系统时做出这项重要发现的。计算机最初连接成网络时，军方也发挥了关键作用，并在其后续发展中继续做出贡献。今天这些固态电子设备难以想象地复杂，可以赋予军用装备各种能力——从最新款的夜视镜到以弹击弹的拦截导弹。21世纪所谓的“网络中心战”的战场上汇集了大量的微型计算机，这些微型计算机又能瞬间对接到拥有超算能力的巨型计算机，使得轮船、飞机、飞船以及它们的有效载荷（包括武器）能在近乎自动的状态下运行。

艾萨克·牛顿在阐述他的地心引力理论时曾假设：一个物体从山顶与地面平行飞出，那么这个与地球大气做相切运动的物体在某个点的动力将正好与地心引力相当。按照他的解释， 96
在离心力和地心引力达成平衡的情况下，这样一个物体会成为地球的卫星。为了验证牛顿的理论，人类花了将近三百年的时间来设计发射装置，而这种装置的军事意义一直以来显而易见。卫星也是非武器军民两用的技术产品，不仅可以被用来从太空观察地球，也可以让自身全部或局部脱轨来打击地面目标。第一颗人造地球卫星在1957年10月4日进入轨道。作为“国际地球物理年”的一个实验项目，它名义上执行的是科学任务，真正的影响却是在军事上。因为按照牛顿的解释，能够把一个物体送入轨道的力量，也可以用来打击位于世界另一端的目标：只要让物体减速，就可以让它落下来攻击预定点位。这个发射装置，按其定义，就是洲际弹道导弹。依据历史上人类将战争范围拓展到海洋和天空的过往经历，防卫专家们立刻做出了“太空军事化”的预言，当然，他们实际上指的是“武器化”。

后来的情况是，几个大国的确将太空军事化了。不过总体而言，对于太空武器化，各国还是有所克制。所谓的近地轨道，包括距离地表一百英里到二万二千英里的同步轨道的范围，这里面充斥各种军事卫星，执行通信、侦察、信号拦截、气象和全球定位等任务。按照1967年签署的《外层空间条约》，美苏两个超级大国以及后来加入条约的世界上主要国家，都承诺不在太空放置大规模杀伤性武器。了解了卫星及轨道技术，大多数国家都意识到，与核武器相比，在绕地轨道放置传统武器并没有多大的意义。因此，尽管卫星已经成为军事行动和地面作战不可或缺的部分，但到目前为止，人类认为还是把武器放在大气层之内

97

为好。

这份非武器性军民两用技术清单还可以很容易地继续添加，比如：用罐头储存食物的技术、履带式车辆、运输机、直升机、

陀螺仪、雷达、GPS，以及操控飞机运动的数字飞控技术等。这些技术所发挥的主流功能和用途，没有发生什么变化。人类自史前时期起，就开始发展非武器性军事技术。很多技术最初出现是为了民用目的，后来才加以改造用于军事，比如前面提到的舍宁根标枪。但有时候，比如计算机和城防技术，它们的军事用途会占据主导地位。在当今世界，有些平民对于使用战争技术会心存芥蒂，即便他们不是用它去杀人或搞破坏。

同样地，军方往往也认为，出于民用目的开发的技术需要进行大量的改造才能满足战场的需求。但更多的情况是，人们不知道某些技术来自哪里，也不明白是什么导致它们产生的。没有几个平民会担心汽车用的发动机技术也会被用于战机、潜艇和坦克；使用电子邮件的人也不会因为这种交流方式源自军方而苦恼（电子邮件最初是美国国防部高级研究计划局的研究人员，为共享研究成果而设计出来的网络信息交流方式）。不仅如此，非武器性技术在现代作战中发挥越来越重要的作用，这表明了一个不断加速的趋势，即冲突会从战场蔓延到现代生活中的百姓社区、交通网络、经济市场、医疗机构和工业领域。

武器性军民两用技术

甚至武器也可以军民两用。社会上的武力设备并不都是军用的。国家在它宣称的控制疆域内拥有（或者声称拥有）垄断性的武装力量，但国家也授权军队之外的公民在特定环境下使用武力，比如警察执法、自卫、安防、狩猎之类。同非武器性技术一样，这些设备在被用于其他领域之前可能各自有其军用或民用的起源。我们同样可以通过几个例子来说清楚这个问题。 98

在军民两用武器中居于显耀地位的，当属舍宁根标枪和它史前的表亲——弓箭。从猎杀动物到猎杀人类，这些技术被证明在战争及和平时期同样重要。当然，狩猎和作战的相似性远不止是在技术方面。两者在进行追踪时，不光需要了解地形、天气和猎杀对象的行为特点，还需要用到情报收集、隐匿行踪、团队合作、通信交流和对敌的勇气。人类不仅是猎杀者，也可能是被猎杀对象，因此需要防卫技术来对付两条腿或四条腿的掠食者。我们所能猜度的是，史前打猎主要战术是“打了就跑”——弱旅在面对强敌时也采用同样的战术。今天我们认为这就是“伏击”，或者是伟大的军事战略家毛泽东所称的“运动战”，其实道理是一样的：趁猎物不注意突然袭击，同时安排好逃跑路线，以便来日再战。因此，最初的军民两用武器是投掷武器，在远距离就能杀伤猎物，同时还让进攻者有机会逃跑。

我们之前讨论过的二轮战车，是另一项军民两用技术。值得注意的一点是，它在军队里兼有武器和非武器的功用。从这个意义上说，它就像船只、飞机、火箭及其他武器平台一样，不仅仅是一件武器，还是一个武器系统的一部分——这个武器系统可以分解为武器和平台。实际上，二轮战车是一个拥有五重用途的技术，它在公元前二千纪的黎凡特地区的战事里曾大显身手，之后还发挥了非作战的用途，比如用于运输、狩猎、赛马和庆典等场合。同船只、飞机和飞船一样，军事版本的二轮马车，把一个移动的平台和上面的武器系统结合在了一起。正是这种基本的配置，使得它天然成为一种军民两用的技术——因为这个平台总有可能被移为他用，就如同船只和飞机也发挥民用功能一样。当二轮战车载着阿喀琉斯去与赫克托决战，就如同龙骑

99

兵骑着马、现代步兵乘坐装甲运兵车投入战斗一样，只发挥了它的运载功能，其功用接近但还算不上是真正的武器系统。而在其他的几项功能上，二轮战车就纯粹是民用的了。在典礼上使用二轮车，无疑是想让人联想起古罗马式的凯旋，从而为获胜者营造出军人般的英勇，但其实与拜占庭帝国时代各个政治派别在君士坦丁堡举行的战车比赛一样，其军事色彩并不浓。

核能是另一项军民两用技术。该技术同时为军方提供武器（炸弹）和非武器（舰船动力）的用途。这项以科学为基础的技术，在20世纪30年代随理论物理和实验物理的迅猛发展而兴起。正是当时在美国的物理学家——其中一些还是逃离纳粹德国的难民，最早提请富兰克林·罗斯福总统重视研发原子弹的可能性。于是，有了第二次世界大战期间紧急开发原子弹的“曼哈顿项目”，并且产生了人类历史上仅有的将原子武器投入实战的结果。此后，核武器继续对战争产生深远的影响，但对作战产生的只是次生的效果——因为这些武器再也不曾为了泄愤而被引爆。相反，它们倒是为“长久的和平”做出了贡献，因为1945年以来大国之间再也未发生热战。同城防工事一样，核技术革命对于那些已发生的战争和未发生的战争都同样重要。1945年以来发生的所有战事，都受到核保护伞的影响。

与此同时，和平利用核能的方式也在不断扩散，其中最主要的用途是在发电和医疗。人们还试图用核能来推动舰船甚至飞机，但大规模的应用目前还只是在军用潜艇和主力战舰上——主要是美国的航空母舰。这项技术第一次投入使用，就摧毁了日本的广岛和长崎，再加上后来几十年里军事和民用领域出现的一些核事故，给核能蒙上了危险的阴影，让人心生畏惧。其

100

实，海军上将里科弗等人的实践早已证明，只要管理严格，核能还是可以安全使用的。

同样让人啼笑皆非的是化学武器在军事和民用上的不同待遇。第一次世界大战期间，德国化学家弗里茨·哈伯把他诺贝尔奖获得者的才能用在研制氯气等致命气体上，这些让人非死即残的可怕手段在战争中迅速崛起。战后，哈伯的辩护词中所谓“无论用哪种方式，死亡就是死亡”的论调，完全忽略了芥子气这样的化学武器所导致的骇人痛苦和残疾。哈伯其实倒是可以像别人那样，用另一套说辞来为化学武器辩护：“化学武器可以不杀死士兵就让他们退出战斗。”如果按此标准的话，芥子气倒是比氯气和光气更为人道，因为它没有那么致命。不过，即便支持毒气战的人能够让世人克服对于这类武器的道德反感，他们仍然面临棘手的武器施放问题。因为无论是使用弹筒还是滤罐——更不要说空投炸弹了，都无法保证所释放的毒气不会随风飘向友军或平民。因此，较之于毒气本身，它的施放才是更大的技术挑战，并因此导致了第一次世界大战后人们再度重申1907年《日内瓦公约》中禁止毒气战的规定。这项禁令在随后的一个世纪里一直发挥着作用，除了几起可怕的例外——而其中大多还是针对平民的。

类似于毒气战的民用方式，主要是用化学物质攻击人的痛觉神经，最常见的是催泪气体和辣椒喷雾。具有讽刺意味的是，催泪气体被《日内瓦公约》归类为化学武器，在作战中是被禁止使用的。可是，大多数国家还是将它用于自己的国民，用来制服罪犯或者控制群体。暴露于这两类物质虽然很少致命，但它们都拥有杀死人的潜力。它们不断被使用表明了一点，民用领域

101

和军用领域的界限在今天的世界正在变得模糊。

另一项军民两用武器技术是炸药。乍一看，这么明显的军用品，似乎不应该归入此类。可是，炸药起源于中国的烟花，其民用功能长期掩盖了我们今天所熟悉的作战用途。所有非核能爆炸物都有同样的物理特性：它们的能量来自化学反应，通过在有限空间内快速燃烧的方式产生。火药，这个最早出现并且最具革命性的爆炸物，在13世纪由蒙古人引入西方，后来又有了经过不断研发而改进的品种。各种火药都是将碳、硫磺和硝石（硝酸钾）组合在一起，其中的诀窍是找到合适的配比——这取决于原料的纯度。到了19世纪，研究者们已经在探索威力更大、体积更小、烟雾更少的新型火药。他们的研究成果包括TNT炸药、无烟火药、火药棉、硝酸甘油、黄炸药以及各种塑胶炸药。这些研发背后多由军方提供资助，但其成果也广泛应用于民间。烟花还在继续给人们带来愉悦，其他的爆炸物也在采矿、民用工程、建筑拆除、雪崩控制和其他建设领域提供帮助。军用的轻武器也为猎人、运动员、治安人员所用，改良的硝化甘油还可以用来治疗一些心脏病。

导弹和火箭不同，但也有重合之处，其混乱的定义往往招致误解。为了更好地讨论这个问题，我们最好把火箭理解为自推动的发射物——由载体内的燃料和氧化剂燃烧后产生的炽热气体向后喷射而驱动。导弹是一种发射物，这个词在这里指的是那些在飞行中主动制导的火箭。自中国人第一次放烟花的时候起，火箭就算是上了天。不过在西方，燃烧助推最初并不是应用于火箭，而是火炮——推进燃料爆燃之后将发射物抛射出去由其自行运动。西方最早的军用火箭出现在18世纪后期，

102

在1812年英军进攻麦克亨利堡时，借由美国人弗朗西斯·斯科特·基写的诗句“拖着红色曳尾的火箭”，其声名流传至今。不过，早期的火箭没有制导，只能大致攻击一个区域，武器的效果有限——这一点直到20世纪中期才得到改变。

这个时期，韦恩赫尔·冯·布劳恩和他的同事将沿弹道轨迹飞行的火箭和原始的惰性导航系统结合起来，研发出的制导V-2火箭可以飞行数百英里，落点的圆概率误差（火箭对该圆形50%的预期命中率）为4.5公里。导弹由此开始发展，后来成为美国和苏联战略军备竞赛的一大基石。即使在今天，它也是大国间和平的一项保证。这些令超级大国达成恐怖平衡的火箭，同样也是太空时代的运载工具。实际上，自1957年苏联人造地球卫星第一次上天，所有飞出大气层的航天器，无论是用固体燃料还是液体燃料，都运用了弹道导弹的技术，其核心技术都是为军事目的而开发的。韦恩赫尔·冯·布劳恩本人就是一个军民相互促进的例子：他最初探索的是民用太空飞行，后来又先后为纳粹国防军和美国军队工作，最后回归到民用目的的阿波罗载人登月项目。“冯·布劳恩范式”在今天还在继续推动并制约着人类的太空飞行。

本书谈及的最后一项军民两用技术是自动武器，或称机关枪。这些武器由单人或多人操作，能自动清理弹仓，通过弹带或弹匣装弹，枪手只需连续击发而不必小心地手填子弹。自第一位火枪手踏足战场起，其成功与否就主要取决于他的射击频率。的确，最早那批火枪手装弹耗时那么长，装弹时只能依赖长矛手保护，因为敌军骑兵会趁射击间隔发动冲击。从17世纪到20世纪，有了一系列提高火枪射击频率的创新：点火装置由火绳变为

103

燧石，弹头和火药被集成到弹药筒，后膛而非枪口装弹，采用带火帽和可取出弹壳的子弹，先是用人力取弹壳、装新弹、上扳机，最后是气动装弹——利用子弹引爆的能量完成装弹过程。从那以后，问题就只是在设计上如何让自动武器更快、更轻和更可靠。美国人在这方面格外擅长，这也许跟他们迷恋持枪权利的国民特性有关。美国人不仅在研发用于军事的机关枪方面一马当先，而且率先把自动武器引入狩猎、体育和安保等领域。在我写作本书时，美国个人持有的枪支总数超过了美国的人口数，也超过了美国军方拥有枪支数量的很多倍。而这些武器最初大多是为军事目的而研发的。人们想不到还有哪种军事技术能够像美国的枪支一样，对民间社会渗透得如此之深了。

那么，我们该如何评价军民两用技术——无论是武器性的还是非武器性的？首先，它们让我们关注一个根本问题：军事研发和生产到底是不是对社会有利？那些奔着作战目的开展的研究有没有对社会发挥补偿性的作用？在一些案例中，这种情况确实存在。不过，我们也得考虑其中丧失的机会成本：如果那些研究者直接将精力投入民用技术研究的话，他们能对社会做出怎样的贡献？循着同样的思路，那些现代化、工业化国家的经济，是否变得需要仰仗政府对于军事研发的支出？信奉“自由企业制度”的主要民主国家，有没有变成如威廉·麦克尼尔所说的“指令经济”——资源都汇集到体现国家意志的项目而饿死自由市场？或者，这些民主国家变成了迈克尔·霍根等历史学家所说的“国家安全至上”的国家？冷战时期的军工联合体，如今在发达国家的控制力有所减弱，但并未消失。不仅如此，我们不清楚现代军事技术，就像战争本身一样，是否会扩散到人类社会的

104

各个领域，从而模糊以前对于军人和平民、战斗人员和非战斗人员、战争与和平的划分？如果军事技术蔓延到平民生活，民用技术也开发军事用途，那么当今世界的军事化程度对于现代社会肌理的渗透之深，会远超我们惯常的理解。通过对军民两用技术的讨论，让我们对这些问题有了更多的关注。

军事革命

20世纪90年代，军方和学界在思考作战技术变革问题时形成了两种分析方式，好似两道彼此相交却相互影响甚微的弧线，如夜间的航船一般彼此擦身而过。但两者轨迹的异同却很好地说明了我们在21世纪之初对于作战技术的理解，同时也凸显了在这个问题上浅尝辄止的潜在风险。通过它们，第二次世界大战以来军事技术的演进方式也得以阐明。

一种分析方式是军事历史学家所描述的军事革命在历史上的作用。历史学家克利福德·罗杰斯认为，18和19世纪所谓的

105

“军事革命”，不过是西方军事评论和分析中的修辞方式。后来，学者们在思考西方改变历史进程的重大变革，特别是发生在美国、法国和俄国以及科学、工业等领域的革命时，尝试以同样的方式思考“军事革命”，这才让这个词在史学界重获关注。1955年，历史学家迈克尔·罗伯茨在他的题为“1560—1660年间的军事革命”的演讲中，特意将两类革命放在一起做了比较。罗伯茨在描述欧洲陆战发生的变化时，认为其背后的推动力量，是单兵火枪和野战火炮在近代（约1500—1789）战场上的运用。这种变化有如下特征：采用了火枪和长矛结合的新战术，战役规模大、时间长，陆军变得更为庞大，以及战事对社会产生了更大

的影响。自文艺复兴时期到法国大革命这段时期的欧洲史学研究，在当时曾被一位历史学家称作“近代的一团乱麻”，而罗伯茨的研究在里面就显得尤为突出。他的研究中强调了瑞典国王古斯塔夫斯·阿道弗斯（1594—1632）所做的贡献，还写了本关于他的传记。

1976年，历史学家杰弗里·帕克在认可罗伯茨的观点之余，对它们做了很大的修改。到了1988年，帕克出版了他具有里程碑意义的著作——《军事革命：军事创新和西方的崛起，1500—1800》，该书对这些观点进行了彻底的改装。罗伯茨的观点至此已经变得难以辨识，唯一没变的是军队规模变大这一点，对此帕克给出的解释是由于“意式要塞”的出现——这是一种为应对攻城火炮而出现的新型城防工程。此外，帕克还为近代军事革命的研究增添了两个全新的要素：一是将时间跨度拓展，涵盖到整个近代时期；二是纳入了欧洲列强在第一波殖民浪潮中的海外扩张。帕克认为，经过这样的扩展，军事革命至少可以为西方的崛起给出部分解释。这种更广义、更有说服力的军事革命，自然可以与符合西方历史标准的政治革命和物理革命比肩而立。尽管帕克的观点在别的历史学家眼里不算意外，但他的书还是在军事史学界掀起了轩然大波，并成为过去五十年里最具影响力的两本书之一——另外一本是约翰·基根的《战斗的面貌》。评论帕克的研究模式成为一股风潮，学者们纷纷找寻其他的例证，再把现象上升为军事革命的理论。历史学家总是能找到自己想要的素材，于是各种关于军事革命的历史研究开始层出不穷。军事革命的历史到处都能找到，比如在中世纪、在亚洲、在美国内战期间、在20世纪伊始的海上军备竞赛当中、在德国的统一战争期间，不一而足。几乎所有这些例证都可以生成自身关于军事革命的定义，对于这个概念是个拓展，但同时也淡化了它的色彩。

与此同时，在由美国军方和国防专家组成的知识群体中，形成了一种不同的学术研究路线，他们称之为“防务革命”。这个概念是由研究防务的美国学者从苏联借用而来的。20世纪50年代的苏联军事分析人员将“军事技术革命”加以理论化，从而提出了这个概念。最初，苏联人的研究重点聚焦在核武器对常规作战的影响上。20世纪60和70年代，随着苏联和美国在常规军事技术上差距越来越大，这种关注变成了不安。因为美国军队在第二次世界大战之后，内部研发热情高涨，技术更迭非常快，让苏联无法跟上。在诸如计算机、高性能飞机、静音潜艇、侦察卫星等高精尖技术领域，美国似乎独步天下，其领先地位可以很快对苏联和所有其他国家形成无可置疑的军事优势。

美国人在读到苏联人的这些研究文章时，也对自身的优势 107 有了新的理解。应不应该重点发展对敌人的非对称优势？苏联人的担心是不是恰恰证明了美国军事研发的有效？

由此，在美国的国防研究者中兴起了支持和强化“防务革命”的运动。精确制导武器，在越南战争失败后获得重点发展，实现了前所未有的精度；围绕未来“电子战场”的讨论越来越多；空军专家约翰·博伊德提出了“观察—判断—决策—行动循环”理论，认为美军可以运用先进技术在观察、判断、决策和进攻这几个环节都做到领先于敌人；目光长远者开始谈论以网络为中心的作战——运用电子网络部队，从而比敌人更快地遂行侦察、通信和战场调度任务。

防务革命，虽然以各种形式出现，但还是存在一些共同特征。没有哪种防务革命是关于核武器作战的——无论是战略核武器还是战术核武器；所有“防务革命”都是谋求美国在常规军事技术上质量领先，针对的对象是苏联/俄罗斯在欧洲数量庞大的陆军。它们都预言，美国的军事实力将达到包括苏联/俄罗斯在内其他国家无法企及的高度。这场运动在20世纪90年代大大加速。第一次海湾战争（1990—1991）中，美军所展示的军事实力，让宣扬“防务革命”的人大为满意。安德鲁·马歇尔，美国国防部净评估办公室的元老，专门对此现象发起了正式研究。比尔·克林顿政府（1993—2001）认为，这倒是一个既可以削减国防开支又不损害美国安全的好办法。确实，防务革命令人难以抗拒的一个诱惑之处就在于，它似乎提供了一个性价比很高的解决方案。

防务革命也引起了前任国防部长唐纳德·拉姆斯菲尔德的关注。2001年当拉姆斯菲尔德回到乔治·W.布什政府的国防部长任上时，他宣布了两项主要目标：部署反导系统（该项目的研发自1983年罗纳德·里根政府宣布后就展开了）和利用“防务革命”来推动军事变革。在拉姆斯菲尔德看来，五角大楼——尤其是陆军，还在固守冷战时期传统战争的思维范式，以为还是在欧洲平原与数量庞大的苏联军队进行战斗。陆军孜孜以求的下一代自行火炮就体现了这种思维模式。当拉姆斯菲尔德再度主政国防部时，名为“十字军战士”的自行火炮项目已经开展有六年之久了。这种口径155毫米的履带式自行火炮可以自动装弹，并能将100磅重的炮弹射到14英里之外。火炮自重43吨，外加40吨的燃料、弹药补给拖车，可以用C-5A和C-17运输机

108

运送到任何拥有3 500英尺长跑道的作战地点。但是，拉姆斯菲尔德认为，陆军想与苏联帝国展开的最终决战是不存在的。他需要的是一支用来打小型战争的精干、灵活的陆军。2001年，“9·11”事件在纽约和华盛顿发生之后不久，他就取消了“十字军战士”自行火炮项目。

拉姆斯菲尔德依赖通过“防务革命”构建的军事能力来应对“9·11”事件。美军在阿富汗精心组织了进攻，动用了为数很少的地面人员，引导对基地组织及其背后的塔利班展开空袭。只花了几个星期的时间，基地组织就被美军的强大火力赶到了巴基斯坦，塔利班也被打得东躲西藏。随后，布什政府把目光转到了伊拉克。国防部长拉姆斯菲尔德的入侵伊拉克的行动没有听从陆军参谋长的建议，而是展开代号为“震慑行动”的大规模空袭，并有大约十五万人的美军部队提供地面支持。美军的巨型战争机器完全碾压了萨达姆·侯赛因的军队——这支军队之前在1990—1991年的“海湾战争”中已经遭受过重创，迫使萨达姆东躲西藏，然后“解放了”这个国家，但随之就面临着一个无休无止的任务：在这个混乱的地方重建一个稳定和公正的国家。109 2003年5月1日，当布什总统登上在波斯湾的美军航母，出现在一个写着“任务胜利完成”的横幅下面时，正是对“防务革命”最好的宣示。美国似乎的确拥有了无法抗拒的军事威力。

但是，之后没有多久，美国的地面部队就暴露出了他们的“阿喀琉斯之踵”。美军新型的“马上作战”方式败给了伏击——他们乘坐的车一辆接一辆地毁于“简易爆炸装置”。这些通过接触、定时或指令引爆的简易炸弹，很快布满了美军车辆通行的道路和桥梁。它们可以简单地通过一部手机引爆，爆

图9　一架F/A–18“超级大黄蜂”正准备从“杜鲁门”号航空母舰的甲板上飞向夜空。这种核动力航母体现了各种系统之整合，正是21世纪到目前为止最为复杂的军事产品

炸物五花八门，有手榴弹、迫击炮弹和大炮炮弹（很多缴获自美军，或来自伊拉克政府崩溃后军队弃置的弹药库），也有大型的未爆弹和自制土炸弹。美国卡车、装甲运兵车甚至坦克对这种武器毫无防备，以惊人的速度陷入战损，车上的司乘人员都遭受了极大的生理和心理创伤。应对这种挑战，美国的“防务革命”可能还需要数年的时间才能生成反制技术。而“基地组织”则在网络上发布了他们运用“简易爆炸装置”攻击美军的宣传影片——这该是又一项军民两用技术，只不过针对的是工业化了的西方国家。

110

这不是拥有高科技、工业化的西方式军队第一次在面对低技术、前工业化、非西方的对手时遭遇挫折。毛泽东在国共内战中，面对蒋介石的西式军队就创造出了他的“人民战争”。伏

图10　“简易爆炸装置”是一种终极反制武器。图中是2003年至2013年伊拉克战争中联军在巴格达缴获的爆炸物。地雷和炮弹被连接到简单的引爆设备上，比如手机，然后被装在联军通行的道路上

111

击战——他称之为“运动战”——在他的策略中占据重要的地位。越南的胡志明，也运用他的战术，夺取了法国人的奠边府要塞，后来又在全国解放的最后阶段打败了美国军队。第二次世界大战以来发生的其他一些战争中，也有装备简陋的派别人员对阵工业强国的正规军队，不少也有类似的意外结果。比如，以色列曾两次镇压巴勒斯坦人的暴动，但在世界舆论场取胜却非常艰难。2001年攻击美国的恐怖分子，在美国本土造成的伤亡超过了1941年日军对珍珠港的袭击，而其使用的武器可以简单如一把裁纸刀。具有讽刺意味的是，这些派别人员和恐怖分子所选择的武器，通常恰恰是当年一经出现即淘汰了野蛮人的武器——火药！

那么，该如何看待21世纪初期军事革命的现状呢？首要的是一种谨慎态度。认为技术甚至高技术就是解决军事问题的灵

丹妙药，既不对路，也很危险。技术当然有助于打胜仗，但并不能保证一定取胜。在研究革命性的军事技术时，历史学家比防务专家做得更好，部分原因是：历史学家进行的是回溯性的研究，而防务专家则是展望未来。尽管历史学家在用到“革命”这个词时也颇为随意，但他们至少明白，“革命”只是在事后才能被定义的，并不是所有的变革都迅猛、剧烈到可以被贴上“革命”的标签。

不仅如此，无论是称其为“防务革命”还是“军事革命”，多少都牵涉到职业发展的手段。历史学家如果声称自己的研究揭示了革命性的变革，就可以凭此扩大学术影响、卖出更多的书，那么，鼓吹“防务革命”的专家则是通过承诺高性价比的变革，进而对决策者施加更大的影响。这并不是说，断言“革命”都是不真诚的，只是表明，“革命”这样的辞藻，无论对于言者还是听者，往往都具有难以抗拒的诱惑力。这两种情形提示我们，在听到谈论“革命”时，我们还是应当保留一份怀疑的态度。[112]

对这两种不同的研究路径进行比较的最后一点，也证明了这些趋势。防务专家常常会援引历史学家关于军事革命的文献，因为借此似乎可以为他们热衷的话题增添学术上的严谨。而研究历史上军事革命的学者则不大会在意防务专家们的言论。历史学家，说到底大多还是学究，终日遨游在弥漫着左倾、反战情绪的学海里。出于理论性和学术尊严的考虑，他们认为自己的研究成果与当下具体的现象之间还是应当保持一定的距离。20世纪90年代到21世纪初，对防务革命和军事革命的研究非常火热，但这股热潮到2010年代就慢慢退去了。[113]

结　语

关于技术和作战的未来，我们能说的或者需要说的只有寥寥数语。技术变革越来越快，已成为20世纪的老生常谈，今天人们也这么认为。但这个陈词滥调背后的核心真相是，变革的步伐确实是在不断加速，而且很有可能继续如此。尤其就军事技术而言，在广泛而刻意的制度化研发之下，更是如此。在本书付梓之际，正向我们走来的技术包括：真正意义上的无人机（无须远程操控），（事先编程的）机器人武器系统，进一步的微小型化的、纳米技术条件下的作战，以及可能最惊人的——自动武器系统（对周边形势能够做出一定程度的独立反应）。我们因此进入的世界，将比人类历史上任何时期都更加危险，但同时又不那么致命。也就是说，作战技术将比以前更加高效，但人口伤亡比例将会比以前减少。我们无法预言这一切到底将会带来什么。

如果说本书提示了答案的话，那么它们也许存在于对术语表所列词汇的理解之中。军事技术将来一定会改变，但是，这些术语背后的原则，比如被人们奉为圭臬的“战争原则”，应该还会继续得到遵循。它们能够超越时空的局限。毫无疑问，来自人类活动其他领域的原则也会影响未来的作战，但这份术语表毕竟为初学者提供了一套工具，也可以为穿越而来的亚历山大大帝提供一个入门读本，用来思考技术与作战的关系。本书的重点之所以放在早期作战，是为了强调：引导技术与作战演进的理念很早就出现了，而且一直以来影响深远。 114

比如，军民两用技术在人类历史各个时期都很发达——从远古的舍宁根标枪到当代的遥控飞行器，那么合理的预期就是：民用技术会继续找到其在军事方面的应用，反之亦然。可以想见，正因为一些技术的军民两用性质，试图限制军事技术转让的努力会遭遇界限模糊的困境。比如，冷战时期对计算机出口的限制，就被证明很难实施。军民两用技术，也反映了历史上一个众所周知的现象，即，军事技术体现的是经济实力。正因如此，经济竞争越来越被视为一种道德所允许的军事实力较量方式。

只要世界还分裂为发达国家和那些没有工业基础设施的国家，这两方之间的武装冲突就注定是不对称的。除此之外，我们很难预言：发达国家还会造出什么样的尖端大杀器？以及那些什么都没有的国家会怎样创新使用低端的攻击手段，比如使用简易爆炸装置、刻意破坏或是盗取大规模杀伤性武器？而发达国家之间，除非有了某种技术突破，否则的话，其对称的武器库倒是可以阻止互相爆发战争。

时下引起广泛关注的网络战，为我们提供了一个分析案例。乍看起来，它对于那些拥有复杂网络的发达国家构成了前所未

有的威胁，它们发现自己很容易受到黑客的攻击。那些新时代

115 “站在门口的野蛮人”，采用的是各种偷偷摸摸和难言抵抗的攻城技术。运用本书提出的一些概念，可以帮助我们厘清这个现象，把它放在历史的情境中加以考量。首先，到目前为止的网络战，无非出于情报获取、蓄意破坏和煽动颠覆这些目的，并非真正的作战。即便是迄今为止最为严重的网络战——2009年和2010年对伊朗核设施的“震网病毒”攻击，也并没有引发战争。网络战运用的是军民两用技术，其攻击的既可以是军方目标，也可以是民间目标。它可以是对称性的（比如，国家主体之间），也可以是非对称性的（发生在国家主体和非国家主体之间）。网络战沿袭的是投掷武器的传统——在远处发动袭击，袭击者有机会逃避直接的报复，这一点使得它更适合较弱的一方。不过，拥有先进网络资源的国家，同样拥有更强大的攻击能力。据说，对伊朗的“震网病毒”攻击，就是美国和以色列发动的。同卫星系统一样，互联网规模庞大，是它易遭攻击的原因之一。据称，朝鲜在“震网病毒”事件之前曾躲过类似网络攻击，就因为他们国内的计算机大多是与互联网隔绝的。所有这些都说明，网络攻击，只不过是数千年来人类需要应对的又一种新技术而已。毫无疑问，网络战在未来的冲突中会占据一席之地，但是强大的国家主体会调用更先进的资源来保护自己，或是用来打击系统的破坏者。网络战最终可能会被归入毒气和反卫星武器那一类，强大的国家之间会避免相互使用，而弱小的国家即便使用也难奏其效。

现代作战的非对称性，似乎有悖于经典理论中战斗双方对于投掷类或打击类武器的选择偏好。历史上，弱小的一方使用

投掷类武器伏击强敌，而强大的一方会尽量靠近对手以图歼灭。虽然很多的弱方战斗人员还是会使用投掷类武器，以便打了就跑，但那些自杀袭击者却选择近距离打击：靠近敌人，杀伤一定范围内的所有人——包括他们自己。历史上曾有过自杀式攻击者——我们马上能想到的就是第二次世界大战时期的日本“神风敢死队”，但这种策略无法持久。原因之一显然就是，这种方法很快就会耗尽人手。与此同时，发达国家却越来越多地使用原本为野蛮人所钟爱的投掷类武器。在现代作战中，包括无人机等空中打击力量在内的这些设备被称为“远射武器”，可以避免动用地面部队造成的伤亡风险。人们现在还在为未来作战开发自动武器系统，所有这些无须将军人置于险地就能打赢战斗的努力，在人类历史上是前所未有的。 116

技术上的相互克制无疑将会继续演化下去——只要其中一方采用了被对手视为威胁的新技术。一个例子就是，在伊拉克和阿富汗发生的简易爆炸装置与装甲运兵车之间的较量，同样，竞相改进的弹道导弹与反导系统也是一对例证。但在后一种情况下，如果真有某个群体想要核攻击美国的一座城市，可能会倾向于采用低端运送平台，比如使用位于纽约东河的一艘船，或者洛杉矶/长滩港口的集装箱。2014年，这两个位于西海岸的港口处理的集装箱，占了进入美国货物的40%，达700万个——其中每一个都可能藏匿炸弹。低技术的“特洛伊木马”战术，较之于高技术的攻城设备，依然是更好的选择。

同样的，巨型化的主张很可能会继续走它的下坡路——正如它在核能时代已经经历的那样。“技术决定论”仍然只会是一个空洞无物的辞藻。因为只要人性不变，人类是有能力驾驭他们的军事技术的；而所谓的“军事革命”——更不用说“防务革命”——依然是罕见的。如果亚历山大大帝穿越回来，他需要学习很多东西，这里所讨论的一些概念倒正好可以为他提供一个

117 起点。

术语表

伏击(也被称为“打了就跑”):一种常常为较弱的一方针对较强的对手所采用的战术,使用投掷类武器。进攻一方多为群体,对猎物突然袭击,在己方安全的情况下尽可能造成对方的损害,然后在敌方的反击或增援到来之前撤退。

适当技术:没有哪种技术是适用一切场景的。要想成功就必须采用适当的技术。也就是说,技术必须适应时空等具体应用场景的要求。以桨帆船为例,它在近岸水域表现良好,但却不适于远洋作战。

非对称技术:冲突情境下双方采用完全不同的作战设备,包括武器和非武器技术。比如,自第二次世界大战以来,航空母舰对于常规主力战舰就享有非对称优势——在进入对方火炮射程范围之前就可以攻击对方。

性能贪欲:历史学家布莱尔·海沃斯用以指称军方热衷于为武器装备添加华而不实的功能。

碳时代:按照战斗中能量来源所划分的第二个时期,大约从

1400年延伸至1945年，处于体能及风能时代与核能时代之间。在碳时代，所有作战中的火力和机械均由内燃机驱动。

骑兵—步兵循环：陆战中，骑兵部队和步兵部队交替占据主

119 导地位。

闭合（参见“锁定”“动能”）：来自科学技术的社会研究的一个术语，意指几种可能技术路径中的一种在发展到占据市场主导地位时，事实上就终结了其他技术路径参与竞争的可能。

联合作战范式：在一定时期的陆战中，所有的战斗人员都采用同样的武器组合方式——即便各国在作战个体使用的武器和作战方式上存在差异。在二轮战车革命后，野战通过骑兵和步兵的组合展开。火药革命后的野战则在此基础上增加了第三支力量——炮兵。

反制技术：一种用来抵消或逆转另一种技术效果的军事技术。

需求拉动（参加“技术推动”）：出于对某种性能需要而推动的技术进步。需求是发明之母。

军民两用：既可以军用又可以民用的技术。

相互克制的技术：彼此相互促进、共同发展的技术，比如城防技术和攻城技术。

巨型化：坚信多就是好，因而拼命往更大、更强的方向发展某种技术。

锁定：经济学术语，意指某种商品的生产者在选定技术上投入巨大（沉没成本），导致无法退回到其他技术路径。参见“闭合”和“动能”。

军事革命：作战方式的改变极其深远和强烈，以至于不仅重新定义了国家之间的武装冲突，而且改变了历史进程；它改变的

不仅是国与国之间的关系，而且是获得强权的方式。本书认为这样的革命有三次：二轮战车革命、火药革命和原子/核武器革命。

投掷类武器（参见"打击类武器"）：无须接近敌人从远处就能实施攻击的武器。也被称为"远射武器"。 120

动能：历史学家托马斯·P.休斯针对"技术决定论"而提出的概念。由于基础设施建设总是围绕某些技术展开，随着时间的推移，结果使得这些技术得以稳固存在，所形成的技术范式人们很难再度改变。美国最初选用轻水反应堆作为核能应用模式，就是一个例子。

非武器性技术：并非直接攻击人或物的作战支持技术。

路径依赖：如果一项技术的发展途径深刻地影响到它最终的形态就被称为"路径依赖"。如果"路径独立"则意味着，无论采取哪种研发路径，一个问题总有一种最佳解决方案。它与"技术决定论"相互呼应。

打了就跑：参见"伏击"。

防务革命：美国在20世纪90年代至21世纪初提出来的军事理论。认为美国在常规军事技术上的进步，尤其在计算机和网络等领域的高科技，将使得美军能够称霸战场。这种理论的热度到2010年代消退了。

打击类武器（参见"投掷类武器"）：需要靠近敌人实施攻击的武器，比如剑、矛和刺刀。海战中的撞击和登船就属于打击类战术。

对称技术：与敌人同步发展的武器或非武器军事技术。

系统集成：多种技术或产品汇集组合，从而形成比各组成部分更强大的能力。拿最基础的蒸汽轮船来说，它需要产生蒸汽的设备、将热能转化成机械能的机器和某种能将机械能转为推力的推进器。

技术天花板：因为缺失一个或多个组成部分而限制了某种技术或系统的发展。比如，真正的潜水艇只有在核能被人类利用之后才出现。

技术决定论：这个标签性辞藻有两个意思：一、技术本身就 121 可以决定历史进程；二、技术与路径无关，最终总会发展出一个最佳解决方案。

技术停滞：技术研发处于静止状态，没有重要进展。

技术推动（参见“需求拉动”）：一种技术能力刺激了大量的应用。例如，蒸汽推力出现以后，船艇性能得以大大改观。

武器平台：装载了武器或武器系统的运载工具。二轮战车、坦克、轮船、飞机和太空船都可以是武器平台。

武器系统：由几种技术或技术产品组成，可用于进攻或防守。所有的武器平台都是武器系统，同样是武器系统的还有骑 122 兵和移动火炮。

索　引

（条目后的数字为原书页码，见本书边码）

A

ABM Treaty《反弹道导弹条约》68
Achilles 阿喀琉斯 16
Adas, Michael 迈克尔·阿达斯 78
Adolphus, Gustavus 古斯塔夫斯·阿道弗斯 106
Advanced Research Projects Agency 高级研究计划局 88，参见 Defense Advanced Research Projects Agency
Afghanistan 阿富汗 2, 3, 69, 109, 117
age of steam 蒸汽时代 52, 55, 71
air superiority 空中优势 60, 85
Alexander the Great 亚历山大大帝 2—3, 115, 117
ambush 伏击 10, 99, 110—111, 116，参见 mobile warfare; pounceand flee
American Civil War 美国内战 52, 54, 74, 107
  CSS Hunley 南部邦联潜艇"汉利"号 53
  Gatling gun 加特林机关枪 72
  technology advantages 技术优势 77
  railroads 铁路 94
appropriate technology 适当技术 54
Archimedes 阿基米德 26
armor 装甲 22, 29—30, 90
  horses 马 29
  infantry 步兵 21, 34
  mounted warriors 骑兵 29—34
  Neo-Assyrians 新亚述人 22—24, 37
  plate 连片护甲 30, 34
  vehicles 运载车辆 100, 110, 117
  warships 战船 52, 71, 73, 77
arms control 军备控制 68, 78—79, 97, 101—102
arms race 军备竞赛 69
  Mediterranean 地中海地区 43
  naval 海上的 48—49, 53
  qualitative 实质的 64
  space race 太空竞赛 65
  World War I 第一次世界大战 59
Arnold Engineering Development Center 阿诺德工程开发中心 87
ARPA，见 Advanced Research Projects Agency，参见 DARPA
arsenal 军械库/军火库 16, 21, 39, 77, 81—82, 84—87, 115
artifacts of technology, defined 技术制品的定义 5—6
artillery 火炮 24, 37, 72—73, 75—76, 79—80, 82, 110, 111
  cannon 大炮 36, 38, 40, 47—48, 51—52, 71, 76, 93—94
  catapult 投石机 23, 25, 26
  field 野（战）的 39, 41, 82, 106, 109
  other throwing engines 其他投射器械 24—25
  siege 围城 36—37, 106
Ashurbanipal 亚述巴尼帕 24
assegai 短标枪 9
Assyrian Empire 亚述帝国 21, 23—24, 27, 36, 43，参见 Neo-Assyrian Empire

asymmetrical warfare 非对称作战 10, 19, 75, 108, 115—116, 参见 symmetrical warfare
Athens 雅典 46
atomic power 原子能 55, 参见 nuclear power
atomic weapons 原子武器 55, 62—63, 81—83, 85, 100, 参见 nuclear weapons
autobahn (Germany) 高速公路（德国）27, 93
autonomous weapon systems 自动武器系统 58, 96, 114

B

B-29（美国空军）B-29 型轰炸机 61—62
barbarian 野蛮人 21, 23, 29, 37, 39—40, 79, 91, 112, 115, 117
  in the Catastrophe “大灾变”时期 18
  chariots 二轮战车 19, 29
  composite recurve bow 复合反曲弓 29
  Enkidu 恩奇都 （《吉尔伽美什史诗》中的人物）13
  Mongols 蒙古人 33, 39
  “the other” 另类 79
Battle of Britain 不列颠之战 63
Battle of New Orleans 新奥尔良战役 74
Battle of the Bulge 阿登战役 84—85
Bismarck, Otto von 奥托 · 冯 · 俾斯麦 78
Bloch, Jan 扬 · 布洛赫 79—80
Bockscar “博克斯卡”（轰炸机）62
Boer War 布尔战争 72
Boulton and Watt 布尔顿–瓦特公司 93
bow and arrow 弓箭 9, 21, 37, 62, 99
  as machine 堪称“机械” 9
  composite, recurve bow 复合反曲弓 21, 29, 33
  crossbow 十字弓 21, 38
  longbow 长弓 34, 36, 38, 79
Boyd, John 约翰 · 博伊德 108
Bronze Age 青铜器时代 13—16, 18
Brunner, Heinrich 海因里希 · 布鲁纳 31, 32
Bush, George W. 乔治 · W. 布什 68, 109, 110
Bush, Vannevar 万尼瓦尔 · 布什 86
buzz bomb 自动飞行炸弹，见 V-1

C

cannon 大炮，见 artillery
capability greed 性能贪欲 64, 89, 90
Carbon Age 碳时代 36, 40, 54—55
Carthage 迦太基 46
Catastrophe, the “大灾变” 18, 20, 23, 30, 39
cavalry-infantry cycle 骑兵—步兵循环 20—21, 30, 32, 35, 38—39, 54
Cervantes, Miguel 米格尔 · 塞万提斯 37—38
chariot 二轮战车 15—21, 29—30, 42, 50, 90, 99—100
cycle 循环 32

heavy 重装 22
scythed 卷镰(战车)21—22
Chase, Kenneth 肯尼斯·蔡斯 40
chemical power 化学战 4, 36—37, 39—40, 69, 102
chemical weapons 化学武器 80, 101
China 中国 17, 27, 33, 40, 91—93, 102, 111
Clausewitz, Carl von 卡尔·冯·克劳塞维茨 5, 95
closure 闭合 57, 参见 lock-in; momentum
cog (ship) 柯克船 48—50
Cold War 冷战 19, 56—57, 65, 67, 82, 89
constraints 制约 115
exaggerated dangers 夸大的危险 88—89
non-weapons technologies 非武器技术 69
paradigm of conventional war 传统战争范式 109
Combined-Arms Paradigms 联合作战范式 4, 20—21, 82
First Combined-Arms Paradigm 第一联合作战范式 20—21, 24, 28—30, 36—37
Second Combined-Arms Paradigm 第二联合作战范式 37, 39, 41, 82
communication 通信 27, 33, 69, 74, 76—77, 82, 90, 95—99
composite recurve bow 复合反曲弓，见 bowand arrow
computer 计算机 3, 63, 81, 86, 95—96, 98, 107, 115—116
Constantinople 君士坦丁堡 23, 37, 46—47, 91, 100
Cortés, Hernando 埃尔南·科尔特斯 75—76
corvus 乌喙 45
counter technology 反制技术 19, 24, 35, 62, 77, 106, 111
Crusader (gun) 十字军战士(自行火炮研发项目)15, 109
cyber warfare 网络战 5, 81, 115—116

D

DARPA, 见 Defense Advanced Research Projects Agency, 参见 ARPA
Davis, R.H.C. 戴维斯 30
Defense Advanced Research Projects Agency 美国国防部高级研究计划局 98, 参见 Advanced Research Projects Agency
demand pull 需求拉动 61, 64, 67, 86, 参见 technology push
Demetrius the Besieger “围城者”德米特里乌斯 23
Department of Defense, U.S. 美国国防部 86—87
Dien Bien Phu 奠边府 111
Dionysius I of Syracuse 叙拉古的狄奥尼修斯一世 26
Don Quixote 堂吉诃德 38
Douglas Aircraft Company 道格拉斯飞机公司 61
Douhet, Giulio 朱利奥·杜黑 60
Dreadnought, HMS 英国皇家海军“无畏”号战舰 52
Dresden firebombing 德累斯顿大轰

炸 37
Dreyse needle gun 德莱塞击针枪 78
dromon 快帆船 46, 47
drone 无人机 63, 114, 117
dual-use technology 军民两用技术 4, 8, 9, 90—105, 111, 115—116
  airplane 飞机 53, 58—59, 61, 64
  bridge building 造桥 27
  chariot 二轮战车 17, 20
  computers 计算机 63
  cyber warfare 网络战 116
  defined 定义 8
  fortification building 城防筑造 11
  ICBM 洲际弹道导弹 64, 66—67
  internal combustion engine 内燃机 53, 55
  nuclear power 核能 55
  prehistoric weapons 史前武器 8—9
  road building 修路技术 27
  Schöningen spears 舍宁根标枪 8
  spacecraft 宇宙飞船 64—66
  steamboat 汽轮 52
  stirrup 马镫 32
  videos 视频 111
duel, warfare as 作战方式的相互克制 3, 50, 60
dueling technologies 相互克制的技术 24—25, 37, 73, 78, 117

E

Ecnomus 埃克诺穆斯 46
Eisenhower, Dwight D. 德怀特 · D. 艾森豪威尔 57, 64—65, 68, 87—89
Egypt 埃及 2, 16, 18, 20
engineering 工程 23—28, 51, 87, 93, 102
Enkidu 恩奇都(《吉尔伽美什史诗》中的人物) 13—14
Enola Gay "艾诺拉 · 盖伊" (轰炸机) 62
Eurasia 欧亚大陆 15, 18, 19, 29, 31, 33, 39—40
Excalibur 亚瑟王的神剑 15

F

feudalism 封建制 31—36
Fort McHenry 麦克亨利堡 103
fortification 堡垒 6, 10—11, 41, 45, 90—93, 98, 100
  ancient 古代的 10—14, 22—24, 26
  classical 古典的 24—26
  Constantinople 君士坦丁堡 37
  Jericho 杰里科 11—12
  Medieval 中世纪的 33, 35, 37
  trace italienne 意式要塞 37, 106
  Uruk 乌鲁克 13—14
Franco-Prussian War 普法战争 78
French Revolution 法国大革命 70, 77, 80, 106
Friendship 7 "友谊 7" 号太空船 66
Fulton, Robert 罗伯特 · 富尔顿 51, 53, 71, 74

G

Gaddis, John Lewis 约翰 · 路易斯 · 加迪斯 89

Geneva Convention of 1907 1907 年的《日内瓦公约》101—102
Genghis Khan 成吉思汗 33
German unification, wars of 德国统一战争 74, 77, 94
gigantism 巨型化 45, 47—48, 52, 54, 73, 116—117
Gilgamesh, Epic of《吉尔伽美什史诗》13—15, 23, 28
gladius hispaniensis 西班牙短剑 28—29
Glenn, John 约翰·格伦 66
Goddard, Robert 罗伯特·戈达德 64
GPS (Global Positioning System) 全球定位系统 69, 98
Great Wall of China 中国的长城 91—92
greaves 护胫套 21
Greece 希腊 2, 9, 17, 21, 24—27, 29, 37, 44
Greek fire 希腊火 46
Gulf War, First 第一次海湾战争 108
gun ports 舷炮舱门 48
gunpowder 火药 23, 33, 55, 112
  changes wrought by 带来的变革 71—75, 102
  revolution 革命 21, 33, 36—41, 47, 69, 71

H

Haber, Fritz 弗里茨·哈伯 101
Hadrian's Wall 哈德良长城 92
Hague Conventions《海牙公约》79
Hannibal 汉尼拔 27—28
Harrison, John 约翰·哈里森 49
Harry S. Truman, USS "哈里·S. 杜鲁门"号航空母舰 110
Headrick, Daniel 丹尼尔·海德里克 75—76
Helmstedt, Germany 德国黑尔姆施泰特市 7
Hephaestus 火神赫菲斯托斯 9, 28
Hiroshima 广岛市 55, 61, 82, 85, 101
Hitler, Adolf 阿道夫·希特勒 85
Ho Chi Minh 胡志明 111
Hogan, Michael 迈克尔·霍根 105
Holland, John Philip 约翰·菲利普·霍兰 53
Homer 荷马 13
Homo sapiens 智人 1, 7—8
Honjo Masamune 本庄正宗(日本名剑)15
hoplites 重装步兵 29
Hughes, Thomas P. 托马斯·P. 休斯 57
Humbaba 芬巴巴(《吉尔伽美什史诗》中的人物)13—14
Hundred Years War 百年战争 30, 33, 34
hunting 狩猎 1, 8—10, 17—18, 20, 99, 102, 104
Hussein, Saddam 萨达姆·侯赛因 109

I

ICBM, 见 intercontinental ballistic missile
IED, 见 improvised explosive device

Iliad《伊利亚特》13, 16
improvised explosive device 简易爆炸装置 10, 110—111, 117
Incas 印加文明 27
industrialization 工业化 39, 64, 68, 70—71, 75—82, 84—85, 88—89, 93—94, 98, 105—106, 111—112, 115
  Civil War, U.S. 美国内战 77
  commercial nuclear power 商业用途的核能 57—58
  industrial revolution 工业革命 70, 94, 106
  military-industrial complex 军工联合体 64, 68, 88—89, 105
  world wars 世界大战 80—85
infantry 步兵 20—22, 29, 30—32, 35, 36, 38—41, 82, 116
  and chariots 与二轮战车 16—18, 21, 22
  dragoons 龙骑兵 100
  heavy 重装 20—21, 29—30
  light 轻装 21
  rate of fire 发射频次 72—73
infantry-cavalry cycle, 见 cavalry-infantry cycle
intercontinental ballistic missile 洲际弹道导弹 56, 65—66, 97
internal combustion engine 内燃机 36, 53—55, 85, 94
International Geophysical Year 国际地球物理年 97
interstate highway system (U.S.) 州际高速公路系统（美国）27, 93
Iran 伊朗 116
Iraq 伊拉克 2, 69, 109—111, 117
Iron Age 铁器时代 18

J

Japan 日本 15, 40, 52, 54, 61—63, 71, 76, 85, 101, 112, 117
jeep 吉普车 16, 100
Jericho 杰里科 11—14, 91
Jomini, Baron Antoine-Henri 安托万-亨利·约米尼男爵 2—3

K

Kadesh, battle of 卡叠什之战 16—17
Kármán, Theodore von 西奥多·冯·卡曼 87
Keegan, John 约翰·基根 107
Kern, Paul Bentley 保罗·本特利·科恩 23
Key, Francis Scott 弗朗西斯·斯科特·基 103
Kilby, Jack 杰克·基尔比 96
knight 骑士 30—35

L

Lepanto 勒班陀 38, 47
Levant 黎凡特地区 2, 10, 15, 17, 18—20, 99
Lindbergh, Charles 查尔斯·林德伯格 61
lock-in 锁定 57, 参见 closure; momentum
logistics 后勤 31, 33, 39, 73, 85, 90

London exposition of 1851 1851 年的伦敦博览会 78
Longbow 长弓，见 bow and arrow
long peace 长久的和平 82—83, 89, 100
long-wave radar 长波雷达 62—63

M

Macedonia 马其顿 2, 21, 23, 29
Maginot Line 马其诺防线 92
Mahan, Alfred Thayer 阿尔弗雷德·塞耶·马汉 46
mail armor 锁子甲 21, 30
Majestic-class battleship 威严级战列舰 71
Manhattan Project 曼哈顿项目 55—56, 85, 100
Marshall, Andrew 安德鲁·马歇尔 108
Martel, Charles 查尔斯·马特尔 32
maryannu 战车武士 16
Masada 马萨达 27
Maxim gun 马克沁机枪 72
McCarthy, Joseph 约瑟夫·麦卡锡 68
McDougall, Walter A. 沃尔特·A. 麦克杜格尔 67
McNeill, William H. 威廉·H. 麦克尼尔 16, 19, 40, 105
mechanization 机械化 72—73, 75, 81, 103—104
Mediterranean Sea 地中海 10, 18, 25—27, 43—47, 49
merchant ships 商船 43, 48, 51—52, 74, 79, 80, 84
Mercury-Atlas 6 水星-阿特拉斯 6 型 66
Merrimac, CSS 南部邦联“梅里马克”号装甲舰 52, 74, 77
Mesolithic period 中石器时期 7, 8
microminiaturization 微小型化 114
military-industrial complex 军工联合体，见 industrialization
military revolution 军事革命 4, 18—19, 36, 105—113, 117，参见 chariot; gunpowder; nuclear
military-technical revolution 军事技术革命 107
Millis, Walter 沃尔特·米里斯 80
missile weapons 投掷武器 9, 37, 45, 48, 63, 65—66, 68, 99, 102—103, 116—117
  arrows 箭 16
  ballistic missile 弹道导弹 25, 56, 65—66, 68, 81—82, 88, 96—97, 109
  incendiaries 燃烧弹 26
  javelins 标枪 21
  longbow 长弓 34
  prehistoric 史前的 9—10
  slings 投石器 21
  snakes 蛇 26
  spears 矛 16, 21
  参见 shock weapons
mobile warfare 运动战 99, 111，参见 ambush; pounce and flee
modernity 现代性 70, 80
momentum 动能 57—58，参见 closure; lock-in; technological determinism
Mongols 蒙古人 33—36, 102
Monitor, USS 美国海军“莫尼特”号

装甲舰 52, 54, 74, 77
Montgolfier brothers 蒙戈尔费埃兄弟 75
Moore School, University of Pennsylvania 宾夕法尼亚大学摩尔学院 86
mounted cycle 骑兵循环，见 cavalry-infantry cycle
Muslims 穆斯林 21, 46, 47, 79
Muwatallis 穆瓦塔利斯 16
Mylae 米拉海战 46
Musashi 日本军舰“武藏”号 52, 54

N

Nagasaki 长崎 55, 61, 82, 85, 101
nanotechnology 纳米技术 114
Napoleon 拿破仑 2, 53, 74, 77, 85
NASA, 见 National Aeronautics and Space Administration, U.S.
National Aeronautics and Space Administration, U.S.(NASA) 美国国家航空航天局 67
National Institutes of Health, U.S. (NIH) 美国国家卫生研究院 86
National Science Foundation, U.S. (NSF) 美国国家科学基金会 86
National Security Act of 1947, U.S. 1947 年的《美国国家安全法案》87
national security state “国家安全至上”的国家 105
Nautilus, USS 美国军舰“鹦鹉螺”号 56
Naval Research Laboratory, U.S. 美国海军研究实验室 87
Naval Sea Systems Command, U.S. 美军海洋系统司令部 87
Nelson, Horatio 霍雷肖·纳尔逊 50—51, 54
Neo-Assyrian Empire 新亚述帝国 22—24, 37
Neolithic period 新石器时代 7, 8
Neolithic Revolution 新石器革命 10—11
net-centric warfare 网络中心战 96, 108
Newton, Isaac 艾萨克·牛顿 96—97
NIH, 见 National Institutes of Health, U.S.
nine-eleven, 见 September 11
non-weapons technologies 非武器技术 12, 69, 73, 90—100, 104
NSF, 见 National Science Foundation, U.S.
Nuclear Age 核能时代 1, 55, 83, 117
nuclear power 核能 4, 55, 58, 90, 100—101
commercial 商用的 57—58, 100
naval ships 军舰 55—56, 58, 88, 94, 101, 110
submarine 潜艇 57, 67
参见 atomic power
nuclear revolution 核能革命 80, 82—84, 100
nuclear weapons 核武器 56, 65, 70, 82—83, 97, 100, 107, 108, 116, 参见 atomic weapons
Nur Mountains 努尔山区 13

O

O'Connell, Robert 罗伯特·奥康奈

尔 40
Office of Naval Research, U.S. 美国海军研究办公室 87
Office of Net Assessment, U.S. 美国净评估办公室 108
Office of Scientific Research and Development, U.S. 美国科学研发办公室 86
Old North River "老北河"号 51—52, 71
Olympias "奥林匹亚斯"号 44
OODA loops "观察—判断—决策—行动循环"理论 108
Otto, Nicolaus 尼古拉斯·奥托 53
Ottoman Empire 奥斯曼帝国 40
Outer Space Treaty of 1967 1967 年的《外层空间条约》68, 97

P

Pakistan 巴基斯坦 2, 109
Parker, Geoffrey 杰弗里·帕克 106—107
path dependence 路径依赖 57—58, 参见 technological determinism
Pax Britannica 不列颠治下的和平时代 50, 71, 77
Pearl Harbor 珍珠港 112
Persians 波斯人 22, 27, 39, 110
phalanx 密集方阵 21, 29—30
Phoenicia 腓尼基 43—44
pilum 短矛 21
piracy 海盗行为 43, 45, 48
poison 毒药 8, 23, 79, 116
Poitiers, battle of 普瓦捷战役 32, 34
Polybius 波利比乌斯 28
polyremes 多层桨帆船 44—45, 54, 参见 trireme
pounce and flee 打了就跑（策略）10, 37, 39, 43, 99, 116, 参见 ambush; mobile warfare
precision bombing 精确轰炸 63, 81
principles of war 战争原则 2, 114—115
Prussia 普鲁士 74, 77—80, 117
psychological warfare 心理战 5, 18, 26, 32, 111
Punic Wars 布匿战争 28, 45

Q

quintuple-use technology 五重用途的技术 20, 99, 117

R

ram, battering 撞击，重击 22, 25
ram, in galley warfare 桨帆船作战中的撞击战术 43—45
Ramses II 拉美西斯二世 16—17
Ramses III 拉美西斯三世 18
RAND Corporation 兰德公司 87
Reagan, Ronald 罗纳德·里根 68, 109
reconnaissance 侦察 59—61, 69, 75, 94, 97, 107
research and development 研发 4, 26, 64, 81, 84—90, 102, 104—105, 108, 114
revolution in military affairs (RMA) 防务革命 107—113, 参见 military-

technical revolution
Rickover, Hyman 海曼·里科弗 55—58, 67, 101
RMA, 见 revolution in military affairs
Roberts, Michael 迈克尔·罗伯茨 106
robotic weapon systems 机器人武器系统 114
Rocket Forces, U.S.S.R. 苏联火箭军 67
rockets 火箭 64—67, 81—82, 85, 99, 102—103, 参见 missiles
Rogers, Clifford J. 克利福德·J. 罗杰斯 36, 105
Rogers, Will 威尔·罗杰斯 69
Roman Republic and Empire 罗马共和国与帝国 9, 11, 17, 21, 24—30, 32, 39, 43, 45—47, 50, 92—93, 100
Roosevelt, Franklin D. 富兰克林·D. 罗斯福 85—86, 100
rostrum 鸟喙状的金属冲角 43
Rumsfeld, Donald 唐纳德·拉姆斯菲尔德 108—109

S

sarissa 重型方阵矛枪 21
Sasanids 萨珊王朝 21, 29
Schöningen spears 舍宁根标枪 7—8, 10, 98—99, 115
science-based engineering 以科学为基础的工程 25, 100
Science: The Endless Frontier《科学：无尽的边疆》86
Science: The Key to Air Superiority《科学：称霸天空的关键》87
scythed chariot 卷镰战车，见 chariot
Scythians 斯基泰人 21, 29
Second Lateran Council 第二次拉特兰会议 79
secret weapons 秘密武器 46, 55, 85
September 11, 2001 2001 年"9·11"事件 109, 112
Shaka Zulu（南非祖鲁族的王）恰卡 9
shield 盾牌 21, 30
Shipping port, PA 宾夕法尼亚州希平港 56
ships of the line 战列舰 49—50, 52, 54
shock weapons 击打武器 9—10, 17—18, 21—22, 29, 31—32, 116, 参见 missile weapons
shortwave radar 短波雷达 63
siege warfare 攻城作战 22—27, 35—37, 41, 45—46, 73, 93, 106, 116, 117
skirmishing 遭遇战 21, 38
SLBM, 见 submarine-launched ballistic missile
sling 投石机 8, 21, 38
Solomon 所罗门 16
spacecraft 太空船 20, 64, 66, 68—69, 88, 96, 99, 103
space race 太空竞赛 65, 67
space warfare 太空战 64—69, 81, 87—88, 97
Spanish Civil War 西班牙内战 60
Sputnik I（苏联发射的）第一颗人造地球卫星 65, 87—88, 97, 103
standoff weapons 远射武器 117
steamboat 汽船 51—53, 71, 73, 76
steam engine 蒸汽机 54—55, 90, 93—94

stirrup 马镫 31—32
Star Wars, 见 Strategic Defense Initiative
Stone Age 石器时代 1, 8, 11
strategic bombing 战略轰炸 60, 62, 80
Strategic Defense Initiative, U.S. 美国的战略防御计划 68
Stuxnet 震网病毒 116
submarines 潜艇 53—58, 67, 77, 84—85, 88, 94, 98, 101, 107
submarine-launched ballistic missile (SLBM) 潜射弹道导弹 56—57
Superfortress 超级空中堡垒(远程轰炸机)62
superweapons 超级武器 16, 18
sword 剑 4, 15, 20—21, 28, 30, 35
symmetrical warfare 对称作战 19, 48, 50, 65, 77—78, 115—116, 参见 asymmetrical warfare
Syracuse 叙拉古 26
Syria 叙利亚 2, 16
system of systems 整合各种系统的系统 42, 71, 110
systems, technological 技术系统 6, 45, 58

T

tank 坦克 5, 20, 54
technological ceiling 技术天花板 36, 48, 50
technological determinism 技术决定论 32, 57, 117, 参见 momentum; path dependence
technological stasis 技术停滞 23, 28, 82
technology, defined 定义技术 5
technology push 技术推动 54, 64, 67, 86, 93, 参见 demand pull
Tell es-Sultan, 见 Jericho
terrorism 恐怖主义 5, 23, 26, 83, 103, 112
terror weapon 恐袭武器 18, 47
thalassocracies 制海权 46, 50
Thor-Jupiter controversy 美国陆军的"托尔–朱庇特之争" 88
Three Mile Island 三哩岛(核电站) 58
Thresher, USS 美国军舰"长尾鲨"号 58
Tokyo 东京 37
total war 总体战 41, 79—83, 94
trace italienne 意式要塞 37, 106
Trafalgar, battle of 特拉法加海战 50—51
transportation 交通运输 17, 20, 61—62, 77, 82, 90, 98—100, 109
triad of strategic weapons 战略武器的三极 56, 65
trireme 有三列桨座的战船 44
Trojan horse 特洛伊木马 23, 117
Troy 特洛伊 16
Tsushima, battle of 对马海战 71
Turkey 土耳其 2, 13

U

Uruk 乌鲁克 13—14, 91
U.S. Army Field Manual 3-0《美军战地手册 3-0》2
U.S. Army Signal Corps 美陆军通信兵 59

V

V-1 (Vergeltungswaffe 1) V-1 型导弹 65
V-2 (Vergeltungswaffe 2) V-2 型导弹 65, 67, 81, 103
Vietnam 越南 108, 111
Victory, HMS 英国皇家海军“胜利”号 50—51, 54
von Braun, Werhner 韦恩赫尔·冯·布劳恩 64—65, 67, 85, 103
von Braun paradigm 冯·布劳恩范式 67, 69, 103
Vulcan 伏尔甘 9, 28

W

walls 城墙 11, 18, 22—24, 26—27, 91—92
  civilization 文明 11
  Constantinople 君士坦丁堡 23, 37
  Demetrius the Besieger “围城者”德米特里乌斯 23
  frontier 前线 91—92
  Jericho 杰里科 11—13
  medieval 中世纪 35—37
  Neo-Assyrian 新亚述时期 23
  Troy 特洛伊 16
  Uruk 乌鲁克 14, 23
war, defined 定义战争 4—5
warfare, defined 定义作战 4—5
War of 1812 1812 年的英美战争 74, 103
Watt, James 詹姆斯·瓦特 93
weapons of mass destruction 大规模杀伤性武器 68, 83, 97, 115
weapon platform 武器平台 16—17, 42, 44, 46—48, 51, 59—64
  B-29 B-29 型轰炸机 62, 64
  boomers 战略潜艇 56
  chariot 二轮战车 16, 20—21, 99—100
  in space 太空领域 64, 68
  low-tech 低技术 117
  reconnaissance 侦察 75
Wright Flyers 莱特飞行器 59
WWI airplane 第一次世界大战时期的飞机 60—61
weapon system 武器系统 17, 30, 36, 50, 63, 65, 89, 99—100, 114, 117
  autonomous 自动的 114, 117
  ballistic missiles 弹道导弹 65
  chariot 二轮战车 17
  naval ship 军舰 50
  platforms (作战)平台 99—100
warplanes 战机 63
weapons of mass destruction 大规模杀伤性武器 69, 83, 97, 115
Weber, Max 马克斯·韦伯 4, 5, 14
Westinghouse Electric Corporation 西屋电气公司 57
White, Lynn, Jr. 小林恩·怀特 31—32
Whitehead, Alfred North 阿尔弗雷德·诺思·怀特海 70—71
Whitney, Eli 伊莱·惠特尼 74
Wilkinson, John 约翰·维尔金森 93
Winner, Langdon 兰登·温纳 58
World War I 第一次世界大战 50, 54, 79—81, 84

1914 1914年 77
airplane 飞机 60
chemical weapons 化学武器 101
Europe's descent into 欧洲陷入战争 59
internal combustion engine 内燃机 94
submarine 潜艇 53
total war 总体战 79—81, 84
World War II 第二次世界大战 4, 37, 39, 73, 80—82, 95, 100, 105, 111, 117
atomic bomb 原子弹 55, 82, 100
Braun, Werhner von 韦恩赫尔·冯·布劳恩 67
demand pull 需求拉动 61—63
missiles 导弹 64—65, 67
nuclear revolution 核能革命 55, 82, 84, 100, 105, 107
research and development 研发 84, 86, 105, 107
rocket research 火箭研究 64
strategic bombing 战略轰炸 67
total war 总体战 80—84
Wright brothers 莱特兄弟 54, 58—59, 71
Wright Field 莱特试验场 87

Yamato 日本军舰“大和”号 54
Yokohama 横滨 62

Alex Roland

WAR AND TECHNOLOGY

A Very Short Introduction

Contents

List of illustrations i

Acknowledgments iii

1 Introduction 1

2 Land warfare 7

3 Naval, air, space, and modern warfare 42

4 Technological change 84

Conclusion 114

Glossary 119

Further reading 123

List of illustrations

1 Schöningen spears **8**
State Agency for Cultural Heritage of Lower Saxony. Photograph by Volker Minkus

2 Walls of Jericho **12**
Photograph by Robert Hoetink, 123RF

3 Pharaoh Ramses II riding a chariot at the battle of Kadesh (c. 1274 BCE) **17**
British Library/Science Photo Library C017/8644

4 Polish army manhandling a primitive cannon at the battle of Orsha (1514) **38**
National Museum in Warsaw

5 The galley *Olympias* entering Tolon, Greece, 1990 **44**
Courtesy of the Trireme Trust

6 HMS *Victory*, the flagship of Horatio Nelson, during the battle of Trafalgar (1805) **51**
National Museums Liverpool

7 American B-29s dropping incendiary bombs over Yokohama, Japan, in May 1945 **62**
Courtesy of the US Air Force

8 The rocket *Mercury-Atlas 6* lifting off from Cape Canaveral Air Force Base in Florida, February 20, 1962 **66**
Courtesy of NASA

9 Aircraft carrier USS *Harry S Truman* **110**
Courtesy of the US Navy

10 An improvised explosive device (IED) in Baghdad during the war in Iraq (2003–2013) **111**
Courtesy of the US Department of Defense

Acknowledgments

The first draft of this book took shape during the spring semester of 2014, when I taught a course entitled Technology and Warfare in the history curriculum at the United States Military Academy at West Point. My students in that course were Taylor Allen, McKenzie Beasley, Jonathan Crucitti, Morgan Dennison, Robert Fee, Jacob Fountain, Lucas Hodge, Bryan Houp, Alex Kukharsky, James O'Keefe, Alexander Reeves, Travis Smith, and Douglas Taylor. I am grateful to all of them for sharpening my thinking on this topic, revealing what worked and did not work, and challenging me at every turn to be clearer and more interesting. This book may not achieve all that they wanted, but it is better for their inputs.

I imposed upon four good friends, distinguished scholars all, to read a draft of this manuscript and share their impressions with me. Daniel Headrick, Wayne Lee, Matthew Moten, and Everett Wheeler all read the manuscript closely and insightfully, offered helpful suggestions, and saved me from many errors of omission and commission that I would have been embarrassed to publish. Wayne Lee was especially helpful in that he had just covered much of the same ground in far greater detail in his masterful *Waging War: Conflict, Culture, and Innovation in World History* (Oxford University Press, 2016). Three anonymous, external reviewers recruited by Oxford University Press provided constructive criticisms and helpful suggestions, which I have attempted to accommodate.

At Oxford University Press, Nancy Toff made clear to me the importance of having an engaged, conscientious, and supportive editor. Her able assistant, Elda Granata, was a model of efficiency, helpfulness, and grace. My copy editor, Ben Sadock, was astute, gentle, proficient, and solicitous.

The index was prepared to my exacting standards with celerity, accuracy, thoughtfulness, and unflagging good humor by my best friend, partner, critic, supporter, and wife, Liz.

I cannot blame any of these good people for the errors and shortcomings that remain.

Chapter 1
Introduction

Humans were born armed. Protohumans had fashioned and used purpose-built weapons before *Homo sapiens* first walked the Earth. These weapons were surely used for hunting and probably for warfare. To make and use weapons and other military technologies is part of what it means to be human. The goal of this book is to trace the coevolution of technology and warfare from the earliest human experience to the present.

Technology and warfare are essentially material. They are communal processes for manipulating the physical world to serve human purposes. Technology seeks to bend the material world in pursuit of human goals. Warfare seeks to bend human behavior by the threat or application of physical force. The two phenomena share a physical and material affinity. A second goal of this book is to trace the evolution of that affinity.

A central thesis runs through the book. Technology has changed warfare more than any other variable. Politics, economics, ideology, culture, strategy, tactics, leadership, philosophy, psychology, and a host of other factors have all shaped warfare. But none of these variables explains the transition from prehistoric to modern warfare as completely as technology. From the Stone Age to the nuclear age, technology has driven the evolution of warfare.

A brief thought experiment might help to crystalize this generalization. Imagine that Alexander the Great came back to life in the second decade of the twenty-first century and found himself assigned to conquer Afghanistan. Might he be up to the task? He conquered that territory in 330 BCE, during one leg of a thirteen-year campaign that took him from his home in Macedonia through what is today Greece, Turkey, Syria, the Levant, Egypt, Iraq, Pakistan, Afghanistan, and beyond. Along the way he met and defeated the best armies of his time, fought through deserts and mountains, carried all the supplies he could not buy or steal along the way, and left relative peace and political stability in his wake. That campaign certifies him as one of the great captains of all time, an obvious master of the art of warfare.

He clearly understood and applied what students of warfare have called the "principles of war." Lists of these principles vary, but all look something like the nine codified in the *U.S. Army Field Manual 3–0* (2011): objective, offensive, mass, economy of force, maneuver, unity of command, security, surprise, and simplicity. These principles are not really rules of warfare, but rather categories of analysis organized as a checklist. Still, experts have viewed them as the keys to success in battle. Antoine-Henri, Baron Jomini (the subordinate and student of Napoleon), said that the "principles are unchangeable; they are independent of the nature of the arms employed, of times and places." If Alexander mastered them in the fourth century BCE, he could surely deploy them to equally good effect in the twenty-first century. In no instance would the principles tell him what to think, but they would always tell him what to think about. There is no reason to believe that he would weigh them any less astutely in the modern world than he did in the ancient world.

No reason, that is, except technology. The one thing that our reborn Alexander would not know and could not learn would be technology. What would he make of explosives, airplanes, satellites, radios, computers, or precision-guided munitions? Citizens of the modern, developed world carry around in their minds tacit knowledge of these technologies, an unconscious understanding of how planes and helicopters remain aloft, why satellites move in orbit, how things blow up, what capabilities reside in the electromagnetic spectrum. Alexander's war in Afghanistan would be over before he could get his mind around such wonders. Everything else about modern warfare would be known or knowable to him. Technology alone would make modern warfare incomprehensibly different from the warfare he knew in his lifetime. As Jomini discerned, the fundamentals of warfare are timeless and immutable. The technology, however, changes incessantly, and transforms warfare in the process. It is the primary driver of change in warfare. It is the variable that would render Alexander impotent. This book will attempt to reveal how and why those changes took the forms they did over the course of human history.

Some arbitrary—but hopefully useful—conventions govern the narrative that follows. First, it is heavily front-loaded. That is, it concentrates on premodern warfare. In that distant past, a set of concepts took root in human practice. One premise of this book is that those concepts—collected in the glossary—offer a key to understanding the kaleidoscopic world of modern military technology. Second, a subordinate thesis highlights one of the most striking, and seemingly contradictory, consequences of changing military technology. While superior technology has generally favored victory throughout history, it has not guaranteed it. "New" and "better" military technologies are not necessarily winners. Technology in warfare does not exist on some absolute scale of effectiveness. Rather, its value is relative to the enemy's capabilities. Think of warfare as a duel, but one in which each party gets to choose his or her weapon. For each side, the choice of weapons will shape the preferred rules of engagement (including no rules at all) and the strategy, tactics, politics, diplomacy, environment, and other conditions of the fight. If, for example, one party chooses a pistol and the other chooses a sword, the outcome is virtually preordained. So too is the likely outcome reversed if the second party chooses a rifle instead of a sword. The technology of the pistol is unchanged, but its relative effectiveness is trumped.

This book also notes that technology and warfare have interacted reciprocally through history. Warfare has changed technology almost as much as technology has changed warfare. This dialectic will be explored within a conventional but slightly simplified chronological periodization, beginning with prehistoric warfare and proceeding through Neolithic, ancient, classical, medieval, early modern, and modern periods. Crossing these basic chronological divides will be periodizations peculiar to the military technologies themselves. One will trace the forms of energy driving military technologies, from muscle and wind to carbon-based chemical reactions to nuclear power. The physical realms in which warfare has been conducted also impose their own chronologies. Land warfare, the oldest and most complex form, has the longest history. Its story is subdivided in a traditional periodization, and further delineated by two "Combined-Arms Paradigms" and two of the three "military revolutions" highlighted in this book. War at sea, in the air, and in space appeared at later times, ending with the convergence of all four realms of warfare in World War II. Finally, there are three topical perspectives on the nature of change in military technology: research and development, dual-use technologies, and military revolutions.

"Warfare," as used in this book, is the conduct of war against an enemy. It is the application or threat of force to kill, capture, or coerce an enemy to do one's will. As such, warfare is generally an activity conducted within a state of war. "War," in Max Weber's classic definition, is organized, armed conflict between states. States are those political entities that claim a monopoly of armed force within their territory. It has become fashionable of late to define war as a condition existing between communities, since so many nonstate actors now appear to be engaging in something like war. But for the purposes of this history, Weber's definition will do. War is a condition; warfare is an activity.

The meaning of "technology" is less clear. In this book, technology is purposeful, human manipulation of the material world. It entails changing some material by the application of power through some tool or machine by some technique. In essence, technology is a process of altering the material world to serve some human purpose. Manipulating ideas, concepts, feelings, relationships, beliefs, emotions, or other human dispositions may be a second-order consequence of some technologies. But it is not technology unless the material world is transformed. Technology, in short, is among the most material of human activities. So is warfare. Indeed, warfare and technology can both shape—even determine—the outcome of war. But they are not war. War may be, as Clausewitz declared, a continuation of politics by other means. But so too is warfare—and its technologies—a continuation of war by other means. Those means are profoundly and inescapably material.

Some activities that fly under the banner of warfare may or may not deserve that title. Cyber warfare, for example, which will be discussed later, certainly uses technology to alter the material world, but it has not yet risen to the level of warfare. Terrorism is not a form of war; it is a technique that may be employed in war or may be an instrument of personal rage or dementia. One may declare war on terrorists, but not on terror. Psychological warfare manipulates ideas more than material; technology may be used but is not essential.

One final definition requires attention. Artifacts of technology are often referred to casually as being themselves technologies. Think of aircraft carriers, tanks, and bombers. These artifacts of technology may be parts of technological systems that sail, shoot, and bomb, but they are not themselves technologies. This distinction is significant for this book, because fortifications and roads are among the most important technological artifacts in the history of warfare. Artifacts such as these, and the technologies that produced them, will appear often in the pages that follow.

The historical record illuminated in this text is primarily Western. This is the history most familiar to the author and richest in evidence. A premise of the book, however, is that the arguments and concepts presented here are universal.

Chapter 2
Land warfare

Prehistoric warfare

We can say very little with confidence about technology and warfare before the dawn of civilization, but we can nonetheless identify some patterns emerging out of the mists of prehistory. One stunning clue arose in the 1990s out of an opencast lignite mine in Helmstedt, Germany. Project Schöningen, named for the mine, unearthed as many as eleven wooden throwing spears that had been preserved for three hundred thousand years in a layer of sediment from a former lake. Spruce stems and pinewood had been crafted into irregular, pointed shafts ranging in length from 5.9 feet to 8.2 feet. Most remarkably, the bodies of the spears are tapered like a modern javelin, weighted forward to fly true. If *Homo heidelbergenses* could throw overhand, they might have launched these spears 35 meters.

These artifacts tell us many important things about prehistoric weapons technology. First, proto-humans were improving upon nature one or two hundred thousand years before *Homo sapiens* emerged. Abundant artifactual evidence has suggested that humans used stones and bones as weapons and that in the Mesolithic and Neolithic periods the stones were being worked artificially into useful shapes. It is reasonable to assume that wood was similarly being crafted, though most of the resulting artifacts

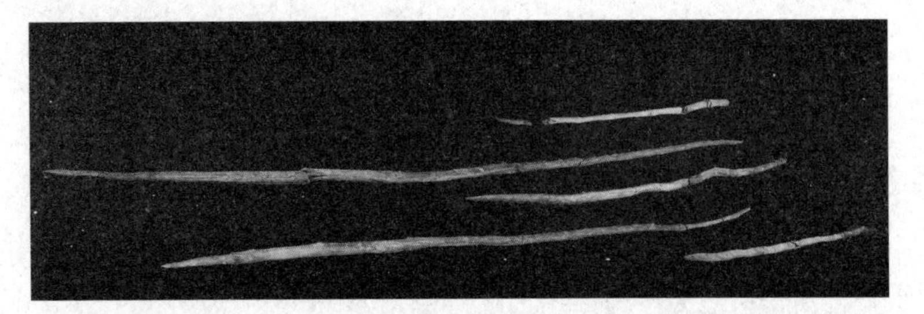

1. The oldest weapon artifacts yet found, the Schöningen spears were dual-use technological artifacts, useful for hunting and warfare. Used by *Homo heidelbergensis* in central Europe three hundred thousand years ago, they prove that humans were born armed.

have long since decomposed. The Schöningen spears establish beyond all doubt that much more sophisticated work was being done on wood much earlier. The spears provide an evidentiary base for inferring wooden spears, pikes (stabbing spears), clubs—even knives—in the intervening millennia. We cannot know for sure if the spears and other weapons were used for hunting or warfare or both, but we can surmise more confidently now than ever before: *Homo sapiens* was born armed.

The next biggest question unanswered by the archeological record is whether Stone Age weapons were for hunting or warfare or both. Most of the remaining reliable evidence—bones, stones, and cave paintings—comes from the late Mesolithic or the Neolithic periods, between roughly twenty thousand and six thousand years ago. By that time, the existing weapons were clearly being used for both hunting and warfare. And there is no reason to believe that a weapon used for one did not find its way to the other. Perhaps poisoned arrows, for which there is some evidence, were reserved for targets that the predator did not intend to eat. But the vast majority of prehistoric weapons—slings, spears, pikes, clubs, knives, axes, maces, atlatls, woomeras—were early instances of what we would now call dual-use technologies. These are technologies that can serve either military or civilian purposes. It is as easy to imagine that some of these weapons were invented for warfare and transferred to hunting as it is to imagine the opposite.

This generalization holds true as well for the greatest of all prehistoric military technologies: the bow and arrow. Invented in the Paleolithic era, more than forty thousand years ago, the bow and arrow has remained in continuous use, in hunting and warfare, up to the present day. While other prehistoric weapons were tools, the bow and arrow is a machine. It has moving component parts and it stores energy. While other prehistoric weapons were intuitive, the bow and arrow required a leap of imagination, an ability to visualize something that did not exist in the natural world. We cannot know if this marvel of creativity developed just once, to be spread by osmosis through the worldwide human community, or if the bow and arrow was reinvented over and over by local geniuses. When the Greeks and Romans conjured a god of arms-making—Hephaestus for the Greeks, Vulcan for the Romans—he was a smith, a metalworker. But the real god of arms-makers was the Paleolithic Edison who invented the bow and arrow.

While many of the secrets of prehistoric weaponry remain hidden to us, the little that we know allows us to make some generalizations about the roots of technology and warfare. First, as already mentioned, most of these technologies of death were probably dual-use. Second, they included both missile and shock weapons, a distinction that continues to the present day. Missile weapons work at a distance, helping to keep the hunter/warrior out of harm's way. Shock weapons—instruments of hitting and stabbing—are more deadly, but they require their wielders to come into contact with their targets. If the target happened to be a large animal—prehistoric man's preferred prey—or another warrior, the encounter could be dangerous. This dilemma runs through all of human history, from Shaka Zulu, "the black Napoleon," who changed the assegai from a throwing spear to a stabbing weapon, to the modern soldiers who mounted bayonets on the ends of their small arms for hand-to-hand fighting when the ammunition ran out.

The contrast between missile and shock weapons also illuminates a third characteristic of prehistoric hunting and warfare. We may infer from studies of nineteenth- and twentieth-century societies fighting with prehistoric weapons that the tactic of choice has usually been pounce and flee. Because big animals and enemy warriors are dangerous, the best way to kill them is by ambush, to attack them by surprise and inflict what damage you can—through missile or shock—and then run like hell. If the attack was successful, you could return later to recover the corpses or dispatch the wounded. If the attack failed, you would live to fight another day. Throughout human history, the ambush has been the preferred technique in asymmetrical warfare, when a relatively weak fighter must attack a stronger foe. In the twenty-first century, the improvised explosive device (IED) has become the new Schöningen spear—the instrument of ambush.

Ancient warfare

The Neolithic Revolution came to the Levant—the land around the eastern Mediterranean—in the tenth millennium BCE, running its course in the region by the middle of the fourth millennium BCE. In those six thousand years, residents of the area learned to domesticate plants and animals and settle in river valleys. They built villages that grew into cities. Some of the early villagers, domesticators of animals, moved out of the river valleys into the surrounding highlands, tending their flocks in a middle ground between the agriculturalists and the hunter-gatherers who continued to roam beyond the pale. All three of these human groupings—foragers, pastoralists, and farmers—developed military technologies for fighting within and between their communities.

The sedentary farmers of the nascent civilizations produced the most important military technology of the ancient world:

fortification-building. While other military technologies helped to determine who won or lost battles and wars, fortification helped to determine if a war or battle would take place at all. As agriculturalists formed sedentary communities to domesticate plants and animals, they found themselves accumulating property beyond the bare necessities. Agricultural surplus, clothing, jewelry, tools of food preparation and consumption, and furniture began to fill up simple shelters. Houses got bigger. Predators—animal and human—roaming the countryside raided these concentrations of food and loot. The simplest defense was wooden poles, set in the ground side by side and strapped together. As mud and stone replaced the wood in these simple houses and walls, new building technologies appeared. Those technologies grew into monumental architecture, the foundation of the city-state. It is a nice question whether the technology evolved first to build the city walls and was then adapted to homes and public buildings or whether the evolutionary arrow pointed in the other direction: perhaps they built the altar and the temple first and then used the same materials and techniques to fortify their enclave. In any event, the dual-use technology of massive, permanent public building became the symbol of the first great civilizations. Indeed, our word for civilization comes from the Roman word for city.

The earliest exemplar is an outlier, geographically and chronologically. Jericho presumably began like other settlements of the Neolithic Revolution, experimenting with the domestication of plants and animals. Unlike the others, however, it fortified its position, an oasis in the Jordan River valley just north of the Dead Sea. By 8000 BCE, a town of perhaps 40,000 square meters, occupied by perhaps two to three thousand people, enjoyed the protection of stone walls 5 feet thick and 12 to 15 feet high. Along one of the walls a tower rose 28 feet above ground level, with an internal staircase that allowed lookouts to climb to the top and survey the countryside for miles around. The Stone Age residents of Jericho left no written record to explain who they were or how and why they came to build such unprecedented defensive works.

2. Rising above the plain north of the Dead Sea, the ancient city of Jericho, today Tell es-Sultan, is one of the earliest exemplars of prehistoric fortified cities, artifacts of non-weapons technologies that shaped war and warfare throughout human history. This photo reveals part of the walls of Jericho and the top of the 12-foot-high tower that once looked out across the plain.

Their location at the intersection of multiple trade routes suggests that travelers might have preyed upon the town, but we cannot know. Archeological research does, however, establish that these walls did not come tumbling down until the sixteenth or fifteen century BCE, though still before the time of the Biblical account. In short, the walls of Jericho worked for more than six thousand years. The settlement seems to have changed inhabitants twice in that time, but not by violent conquest.

The ancient city-state of Uruk, on the Euphrates River, offers far more insight than Jericho into early monumental fortification. Uruk flourished around 2900 BCE, in the middle of the Bronze Age. Modern archeology has provided reliable artifactual evidence about this truly monumental city, which we can correlate with a written foundation myth. These two sources of information illuminate both the technology of monumental building and the role it played in society.

The *Epic of Gilgamesh*, like Homer's *Iliad*, circulated as oral tradition until it was captured in various written forms by authors with no firsthand experience of the events they reported. Decades of scholarly study have yielded widespread consensus on the original myth and produced volumes of speculation on what it might mean. We know with a high degree of certainty that Gilgamesh, a real person, ruled the Mesopotamian kingdom of Uruk in the first half of the third millennium BCE. The discerning reader must sift the rest of his story to separate the wheat of historical plausibility from the chaff of apocryphal legend, but even legend is instructive. Two-thirds god and one-third man, Gilgamesh ruled Uruk for 126 years. His epic tells of a series of heroic quests in pursuit of fame and immortality. One quest takes him to the cedar forest ruled by Humbaba, a fallen god who also controlled the underground river that led to hell. This cedar forest was most likely in the Nur Mountains of modern Turkey, on the upper Euphrates River. Accompanying Gilgamesh on his quest is Enkidu, a barbarian seduced by civilization. When they meet

Humbaba, Gilgamesh slays him with magic weapons forged for him by a good deity, but not before Enkidu falls victim to Humbaba's deadly gaze. Gilgamesh travels to hell and back in search of immortality, but returns knowing that he will eventually follow Enkidu to the grave.

While the *Epic of Gilgamesh* explores love, life, and death, it also sheds light on the more material world of Bronze Age Mesopotamia. Gilgamesh went to the cedar forest to get wood for the city gates of Uruk and perhaps also for the ovens that baked the mud clay bricks with which Uruk and its walls were built. Gilgamesh boasts often of the hard-baked bricks of his city, a luxury available only to those with the wealth and courage to acquire the necessary firewood. In Gilgamesh's day, his city boasted a wall about 3.4 miles around, enclosing 2.3 square miles, roughly 220 times the size of Jericho. Its population surpassed eighty thousand in Gilgamesh's time, making it the largest city in the world. A moat surrounding the city provided an extra barrier against invaders and threatened to drown those who might attempt to dig under the city walls. Within the 25-foot-thick walls, the city had its share of monumental civilian architecture, temples and other public buildings designed both to house community functions and to awe the observer. Gilgamesh cherished above all else the title of "Builder of Walls." He was master of the technology that guaranteed the security and prosperity of his city. Even more than the military prowess afforded by his magic weapons, Gilgamesh staked his reputation on the fortifications he built, "Uruk of the strong walls." By this time, of course, prehistoric warfare had clearly turned into Max Weber's organized armed conflict between states.

While monumental fortification was transforming conflict around other urban centers such as Babylon and Nineveh, two other military technologies were changing field warfare. The metal that gave the Bronze Age its name soon replaced the stones in arrowheads, spears, knives, and other instruments designed to penetrate human and animal flesh. Furthermore, bronze made possible an entirely new weapon: the sword. Earlier stabbing weapons were limited in length by the weight and brittleness of stones and bones. Bronze—a mixture of copper and tin—could be worked into blades of several feet in length, edged on both sides and sharpened to facilitate stabbing and cutting. The very earliest writings from the new civilizations reveal apparently well-developed foundation myths in which heroes with godlike qualities deploy weapons endowed by the gods with supernatural powers. Gilgamesh, for example, carried a bow of prized Anshan wood and an ax—"the Might of Heroes"—fashioned for him by gods. His weapons, we are told, weighed 600 pounds, obviously instruments that only the demigod Gilgamesh could lift, let alone wield to effect. More than any other military technology around the world, the sword quickly took on symbolic meaning in legend, folklore, and mythology. From King Arthur's Excalibur to the very real Japanese Honjo Masamune, swords have romanticized warfare more than any other technology. Soldiers around the world still wear them in dress parades as reminders of a time when warriors believed—or wanted to believe—that certain instruments of warfare could transmute virtue, justice, honor, or even godliness into military victory. Or perhaps it was the other way around; perhaps magical swords were a sign of a god's grace. Modern warriors still give their weapons names like Zeus, Patriot, Crusader, and Peacemaker.

The greatest of all Bronze Age weapons did not arise from the new civilizations, nor was it made of bronze. The wooden chariot evolved on the Eurasian steppe and burst into the Levant in the seventeenth century BCE, unbidden and unanticipated. There had been war wagons in Mesopotamia in the fourth millennium BCE, but these heavy trucks, riding on four fixed, solid wheels and pulled by asses or oxen, likely carried warriors and their equipment at a lumbering pace to the site of battle. The chariot, in contrast, sped across the battlefield on two spoked wheels behind two or four horses, faster than soldiers could get out of their way. They appear to have swept all before them, running circles around or through infantry armies and forcing their enemies to surrender or arm themselves in kind. For almost six hundred years, these "superweapons" of the ancient world, as historian William McNeill has called them, forced the great powers or would-be great powers into an unprecedented arms race. To compete, states had to develop fine woodworking skills in territories with little wood, amass horses in horseless regions, and build arsenals, stables, and repair facilities for both home defense and operations abroad. So great was the demand for these exotic machines of war that an international mercenary class appeared—the *maryannu*—to sell their services and equipment to those states that could not master the technology or afford their own standing fleet. King Solomon reportedly amassed 1,400 chariots.

By some accounts, the greatest chariot battle of all time occurred at the beginning of the thirteenth century BCE, when the Egyptian chariot forces of King Ramses II of Egypt met the chariots of the Hittite king Muwatallis, outside the city of Kadesh, on the Orontes River in modern Syria. Mystery and controversy surround the battle, but there is little doubt that a campaign involving thousands of chariots and tens of thousands of soldiers reached a climax that imperiled Ramses and forced his withdrawal to Egypt. The decisive battle of the era went to the quasi-civilized Hittites.

Remarkably, for one of the most important weapons in all of world history, we do not know for certain how the chariot was used in battle. Majority opinion favors a weapon platform. One charioteer drove the vehicle into range of the enemy forces while one or two passengers shot missile weapons—arrows or spears—at the enemy formation. Another candidate is the jeep, a vehicle that takes warrior elites to the site of the combat, where they dismount to fight on foot. This use is portrayed in the *Iliad*, at the end of the Bronze Age, when Achilles, for example, is carried to the walls of Troy to call out Hector for hand-to-hand combat. When Achilles slays the Trojan champion, he drags the body around the city walls

3. The chariot revolutionized warfare in the second millennium BCE. This depiction of the pharaoh Ramses II at the battle of Kadesh (c. 1274 BCE) shows him riding over the bodies of Hittite soldiers slain by his arrows. The chariot, a dual-use technology, was the world's first weapon platform on land.

behind his chariot. The third possible use for chariots is shock: driving the vehicle directly into the enemy infantry formations with the archers and spearmen on board firing into the ranks as the chariot passes through.

However it was used, the chariot receded from the Levant even more quickly than it had appeared. After about 1200 BCE, the chariot lost its dominance of the Levantine battlefield, never to be recovered. The technology migrated both east and west, seeing use in India, China, Greece, Rome, mainland Europe, and even England and Ireland in succeeding centuries. But eventually it disappeared from these battlefields as well, retiring to uses of hunting, ceremony, transportation, and sport, such as racing. What could have caused such a dramatic and rapid eclipse of such a powerful weapon system? Because 1200 BCE was about the time when the Bronze Age gave way to the Iron Age, some scholars have conjectured that new iron weapons allowed infantry to stand up to chariots. But this interpretation has fallen out of favor. Alternatively, the explanation might be economic—that the chariot arms race was ruinously expensive on all the participants, finally exhausting their ability to finance it. Others have identified an event known as "the Catastrophe" as the agent of change. Around 1200 BCE, waves of barbarian warriors from the Eurasian steppes, impelled perhaps by environmental or climatological forces, moved into the lands of southwest Asia surrounding the Black Sea, the Aegean Sea, and the eastern Mediterranean. As they advanced, they drove before them the residents of these regions, who in turn fell upon their neighbors to the south, creating a cascade of forced migrations and invasions that climaxed in the waves of "Sea Peoples" that washed ashore in Egypt in the thirteenth century BCE. Ramses III met these amphibious invaders in his chariot, something of a last hurrah for this superweapon in the Levant.

One explanation for the eclipse of the chariot that seems to fit with all the evidence is the possibility of new infantry tactics, perhaps introduced by the steppe warriors who launched the Catastrophe. Being horsemen themselves, they perhaps knew that horses would not charge into a wall or a solid line of men holding their positions. If the chariot really was being used in shock, and if the infantry armies of the day learned that they could stop them by simply holding their ground—perhaps stiffened in their resolve by new iron weapons—then it might be that the chariot suddenly lost its menace. Perhaps it was a terror weapon all along, exerting more of a psychological than a material impact.

In any event, the chariot lost its preeminence in Levantine warfare, retiring to supporting roles behind the lines, on the roads or the parade ground, or in the circus or on the hunt. But the chariot's brief, dramatic reign over Western combat effected a military revolution, the first of three that will be highlighted in this book. "Military revolutions," as used here, are transformations of warfare so profound and sweeping that they not only redefine the nature of warfare but also change the course of history by shifting the relationship between states and access to coercive power. As William McNeill said of the chariot, it "transformed the entire social balance of Eurasia." While bringing about such a transition, the chariot also introduced a number of issues that were to recur often in the history of technology and warfare.

First, it was a truly revolutionary weapon. It forced all states within its reach to adopt it, counter it, or make peace with its masters. In the terminology of this book, the fighting options were for symmetrical or asymmetrical warfare. The asymmetrical option required a counter technology or technique. For six hundred years, no one appears to have been able to stop it. Instead, they adopted it—one measure of a true revolution. That is, they chose symmetrical warfare, like the Cold War arms race of the twentieth century. Second, chariots were invented not by civilizations but by barbarians. The whole history of innovation in military technology has been centered on civilization and has usually given civilized states leverage over the primitive. In this case, however, the barbarians of the Eurasian steppe initiated the revolution—first by domesticating horses and then by harnessing them to combat vehicles. Once the civilizations encountered this new technology, they all embraced it or succumbed to it.

And that is the third point. Until 1200 BCE, when the chariot's impact waned, there seems to have been no counter technology. Maybe some armies sought terrain on which the chariot could not function, but there appears to have been no anti-chariot. Chariot fleets fought chariot fleets symmetrically. Fourth, this technology diffused with great comparative speed, from the steppe to the Levant and then to most of civilized Eurasia. The chariot seems to have instilled in military leaders an imperative to adopt or surrender. Fifth, this technology was arrested either by its own inherent limitations or by innovative countermeasures. Either the chariot collapsed because of its own cost or warriors developed a technique to stop it.

Sixth, like the sword, the chariot took on transcultural symbolic value. Egyptian pharaohs sought to be portrayed in their chariots, either on the hunt or in war. The chariot became the vehicle of choice for civilian or military leaders who wanted to create a popular aura of command, power, and triumph. Seventh, the chariot began a cavalry-infantry cycle that has persisted into the twenty-first century. Through long epics of recorded history, especially in the West, warfare has been dominated alternatively by mounted or infantry warriors, each employing alternating weapons systems that commanded the battlefields of their day. The chariot roaring down from the steppes launched the first mounted cycle; the eclipse of the chariot returned the Levant to an infantry cycle. In the pages that follow, this book will explore and try to explain the forces—especially technology—that moved history from one of those cycles to the next. Eighth, along the same lines, the chariot became not just a dual-use technology but rather a quintuple-use technology: it was used for war, transportation, hunting, ceremony, and sport. The steppe nomads probably developed it for hunting; others found different uses. Ninth, and finally, the chariot was the first ground-warfare weapon platform. Nothing on a par with it would appear in land warfare before the tank in the twentieth century. But in its social and military function, it was the forerunner of the naval ship, the military airplane, and the spacecraft. Like those platforms in other realms, the chariot required one part of the crew to operate the vehicle and one part to operate the weapons. It was way ahead of its time.

The first combined-arms paradigm

The Catastrophe, c. 1200 BCE, plunged the Levant into a "dark age" of economic, political, military, and technological stagnation. The eclipse of the chariot left land warfare in a Combined-Arms

Paradigm unrelieved by military innovation. Until late in the Middle Ages, field armies of the civilized states built their military power around a phalanx of foot soldiers, supported by mounted warriors and light infantry auxiliaries. For a thousand years, all of these soldiers deployed basically the same arsenals. The heavy infantry of the phalanx carried spears and swords. The spears ranged from missile weapons like the Roman *pilum* (really a javelin) to the Macedonian *sarissa*—a heavy pike more than 20 feet long. Swords varied from the short, stabbing Roman *gladius* to the long, slashing swords of the Sasanids. A variety of other stabbing and clubbing weapons also might be carried for close-in fighting. These heavy infantry were armored as well. All carried a shield supplemented by various kinds of body armor. Most wore helmets and some sort of breastplate or mail, complemented perhaps by greaves (to protect the shins from kicking in hand-to-hand combat) and other specialty guards.

Light infantry might support the heavy by providing missile barrages from the flank or front (skirmishing). They usually shot bows and arrows, javelins, or slings. Because they relied on mobility for protection, they wore far less armor—if any. Mounted warriors rode chariots, horses, or camels. The few scythed chariots documented in this time were clearly used for shock, riding into enemy infantry formations, but chariots may also have been used as missile platforms and for flanking, screening, and scouting. Cavalry performed the same functions.

Infantry dominated the Combined-Arms Paradigm until the waning days of the Roman Empire. Then a new cavalry cycle began to emerge, lasting until the gunpowder revolution. Assyrians, Persians, Greeks, Macedonians, Romans, Scythians, barbarians from thc forests and the steppes, and Muslims from the desert fought differently from one another, but the differences were in organization, tactics, strategy, and culture. Through both infantry and cavalry cycles, the technology of classical and medieval land warfare remained fundamentally the same, locked in a static paradigm of field combat. All states adapted the existing repertoire of arms and armor to suit their budgets, their natural resources, their labor and conscript pools, and their ways of war.

The Neo-Assyrian Empire

One apparent exception to this pattern illuminates the larger phenomenon. The Neo-Assyrian Empire (911–612 BCE) was a fully formed militarized state, the first recorded predator state in world history. Over the course of three centuries, the Neo-Assyrians set the world standard for continuous, rapacious, remorseless, and expansionist warfare, distinguished by self-conscious innovation in field and especially siege warfare. They built warships while still a landlocked state. They are depicted fording rivers in combat gear floated by air-filled animal bladders. They built roads to connect the ever-expanding bounds of their empire. And they armed and equipped their soldiers with the most modern, high-quality uniforms, armor, and weapons they could produce. They resurrected the chariot as a heavier vehicle, riding on thick wheels with as many as a dozen spokes. This new machine allowed them to carry a crew of four and perhaps even ride into the rough terrain of the foothills and mountains around the Mesopotamian valley. The heavier chariots may also have facilitated shock tactics, driving directly into enemy infantry ranks. There is even some evidence that the Assyrians pioneered the introduction of the scythed chariot, designed to cut down enemy foot soldiers in formation. All of these chariot innovations suited the aggressive, bloody, horrific style of war practiced by the Neo-Assyrians.

Neo-Assyrian innovations in siege warfare proved even more dramatic. To the traditional tool for attacking fortifications—ladders—they added siege towers on wheels, giving their soldiers direct access to the defenders atop the city walls. They introduced a kind of ram for battering city gates and another for the walls themselves. These wall rams came in two versions, one to batter the walls and another to pick at the mud clay bricks of most

Mesopotamian fortifications. The Assyrians even dug mines beneath city walls and perhaps built catapults to fire over them.

In the end, however, the Neo-Assyrian florescence in military technology failed to end the technological stasis that settled on Western warfare—both field and siege—after the Catastrophe. Not even their ingenious siege devices could alter the preponderance of power enjoyed by sophisticated fortifications. Armies at the gates of robust fortifications could surround and starve out the population within the fortress. They could poison the water supply. They could slaughter the inhabitants of conquered cities in order to terrorize the defenders of other fortresses into capitulation. Or they could gain entry by betrayal or trick—as the Achaeans did with the Trojan Horse. Though Sargon II (r. 721–705 BCE) and other Assyrian kings fancied themselves destroyers of walls—perhaps a counter to Gilgamesh's pride in being a builder of walls—there is little evidence that their siege technology conquered many cities. The greatest destroyer of walls in the ancient world, "Demetrius the Besieger" of Macedonia, failed to take Rhodes in a year-long siege in 305–304 BCE that fielded the largest siege engine of the age, a nine-story mobile siege tower bristling with catapults at multiple levels. Even that engineering marvel, however, succumbed to Rhodian assault and defensive catapults. As historian Paul Bentley Kern observed of this campaign, "Ancient siege warfare had reached a technological dead end that was not escaped until the introduction of gunpowder a millennium and a half later." Most urban conquests by the Assyrians and their successors followed time-honored traditions of barbarians, weaker powers, and even great empires down to the Middle Ages—surround and squeeze. Siege technology would not upset this balance until the walls of Constantinople fell in 1453.

Still, there is no gainsaying the inventiveness and the effectiveness of the Neo-Assyrian Empire. What could account for this effervescence of technological innovation? Were the Neo-Assyrians simply more intellectually curious than their contemporaries—a hypothesis reinforced by the unparalleled library of their king Ashurbanipal (668–627 BCE)? Was their population smaller than their ambition, sending them in search of labor-saving machinery that might leverage their military power? Or are all militaristic states ever vigilant in the pursuit of new arms and equipment? Whatever the reason, the Neo-Assyrians introduced a raft of new military technologies.

These innovations did not, however, ensure success. Rather, they introduced a pattern of dueling technologies in siege warfare that continued unabated into the modern world. Defenders build strong walls. Attackers develop a siege tower to scale the walls. The first side sets the siege towers on fire. The second side covers its siege towers with wet animal skins to retard the flames. One side develops siege artillery to breach the walls, and the other side emplaces comparable machines on its walls to shoot at the besieger's machines. And so it goes. More than simply counter technology, dueling technologies entail an ongoing, machine-like, reciprocal pattern of innovation. Through most of the First Combined-Arms Paradigm, defensive fortifications succeeded more often than sieges, Neo-Assyrian boasts to the contrary notwithstanding.

Classical warfare

Shortly after the fall of the Assyrian Empire in 612 BCE, Western civilization left what I call its ancient period (3500 to 500 BCE) and entered the classical era (roughly 500 BCE to 500 CE), the age of Greece and Rome. Still operating within the First Combined-Arms Paradigm, the Greeks and Romans improved upon the siege engines and other military technologies of the Assyrians, and they also refined written records, bureaucracies, roads, and fortifications. In the process they pioneered engineering in a decidedly modern form. It is entirely possible that the Assyrians had their own military engineers; the images and artifacts they left behind suggest as much. But only with the Greeks and Romans does the literary record confirm such engineering carried to high levels of refinement.

As with so much in Western civilization, this story begins in Greece in the middle centuries of the first millennium BCE. Residents of the Greek city-states began to fashion a civilization more prone than its contemporaries to interpreting the natural world rationally and to cultivating philosophy, science, politics, culture, and art as ornaments of the state. Some students of Western civilization have found its roots in what they call the "Greek miracle." In the military realm, one historian has gone so far as to claim that the classical Greeks even invented a "Western way of war." Most scholars find that assertion unconvincing, but there is nonetheless widespread agreement that classical Greek civilization introduced the world to many of the concepts, beliefs, and patterns of thought and feeling that have come to make up the Western worldview.

In the realm of military technology, the most important Greek contribution was what we would now call science-based engineering—that is, the design, construction, and use of machines and structures based on mathematics and, in modern parlance, "science." Hellenistic Greeks proved especially adept at siege technology and its reciprocal, its dueling twin, the refinement of fortification. Greek ideas and engines spread around the Mediterranean world, along with the rest of their cultural legacy, taking root most spectacularly in the Roman Republic and Empire. There, military technology achieved a transcendent importance, in many ways outshining the field warfare of the vaunted Roman army.

Between them, the Greeks and Romans bequeathed to the world a panoply of siege engines. In addition to rams and mobile armed towers, they developed multiple forms of artillery: catapults, ballistae, onagers, scorpions, etc. These latter throwing engines, forerunners of modern artillery, all stored and released energy from one of three sources: tension, torsion, and gravity. (Tension and torsion machines, respectively, stretched or twisted organic materials such as rope, wood, or animal hair or sinew.) The hurling machines, which launched their projectiles in curving ballistic arcs, might have thrown incendiaries, animal carcasses, snakes, or other unpleasant missiles into the enemy fortress, but it is hard to imagine them doing much harm to the walls. The direct-fire weapons might have chipped away at the walls and perhaps opened some breaches around gates or other weak points, but they lacked the power to readily force a breach even in the mud brick walls of Mesopotamian fortifications. Towers to go over the walls and mines to collapse them held out more promise, but moats and fire could limit their effectiveness. It is entirely possible that the major impact of these ingenious machines was psychological. It is likely that nontechnological means remained the most effective forms of siege warfare—negotiation, starvation, terror, ruse, and betrayal.

Still, the military engineers of the classical world made many other contributions besides clever siege engines. First of all, they instilled in kings the belief that technology could deliver military advantage. Some engineers held positions at court, and others moved about the Mediterranean selling their services. Dionysius I of Syracuse went so far as to establish a center for military research and development, producing, by one account, the catapult. Archimedes, the greatest mathematician of his age, died in the futile defense of Syracuse against Roman attack, but not before inventing a system of mirrors to burn an invading fleet with concentrated sun rays. He may even have invented a highly leveraged crane to capsize enemy ships, though it is always hard to know how much of the classical war stories to believe. His contrivances and those of his Greek colleagues may have been more clever than effective.

The engineers' contributions to what we would now call civil engineering were even more impressive and verifiable. The Romans laid down 55,000 miles of primary and secondary paved roads that circled the Mediterranean and sped the Roman legions to duty assignments around the empire. Brilliantly engineered, these roads adapted standard plans to local materials and terrain, producing remarkably straight roads of various depths, widths, and cohesion along elevation contours that minimized rising or falling grades. Bridges of wood and stone complemented the roads. A classic dual-use technological artifact, the Roman roads served the military and strategic goals of the state, while promoting government, commercial, and personal travel that bound the empire together. The Persians, Assyrians, and Chinese built comparable state roads for commerce, war, and government communication, as did the Incas and others in later centuries, but not until the German autobahn and the US interstate highway system did any state road network rival that of Rome.

Soldiers of the Roman army built many of those main roads and applied the same skills and knowledge on campaign. The army forded rivers with pontoon bridges, and Caesar twice built wooden bridges across the Rhine in the face of the Germanic tribes, demonstrating that resistance to Rome was futile. The same mentality fueled the Roman army's preferred siege technology, earthen ramps built up to the top of the enemy's wall. This was an ancient siege technique raised to high art by the Romans. The Roman ramp at Masada is still in place, and the footprint of one of the camps that protected the army while it was under construction is clearly visible nearby. Roman republican and imperial armies lost more than their share of battles, often due to poor leadership. Hannibal, for example, dealt them some of their most devastating defeats on their home territory. But always the Romans came back, fighting relentlessly and doggedly until they won. It may be said that they won more victories by engineering than by fighting. For them, military engineering was not just an instrument of state power; it was an ethos, a way of war. Eschewing the elegant, mathematics-based engineering of the Greeks, the Romans embraced a pragmatic, cut-and-try engineering, much of it no doubt learned on campaign and passed on in doctrine. Many enemies came to terms with Rome not because they were defeated but because they were weary.

In spite of Rome's achievements in engineering, however, field warfare in the classical period remained bound within the technological stasis of the First Combined-Arms Paradigm. The most remarkable innovations were simply variations on static technological forms. The *gladius hispaniensis* is a case in point. The gladius was the generic sword of the Roman legionnaire, shorter than most swords before and since but also varying greatly over time in length, blade style, hilt, and especially material. Its history illustrates a penchant for assimilating innovations in military technology. The Romans were unsurpassed, said Polybius, "in the readiness to adopt new fashions from other people, and to imitate what they see is better in others than themselves." The Romans already had a short sword of wrought iron when they discovered the special qualities of Iberian swords during the Second Punic War. Hannibal Barca launched his invasion of Rome from his family's Carthaginian colony on the Iberian Peninsula. The Romans soon discovered that the swords from Spain were stronger than theirs and held their point and edge much longer. They could, in fact, be honed to a razor edge, magnifying their effectiveness and durability in battle. The Romans studied the Iberian technique of sword-making and took it home. They did not, however, take home the Iberian iron ore that gave Toledo steel many of its special qualities. Seldom, therefore, did Roman knockoffs achieve the attributes of the authentic *gladius hispaniensis*. In time the Romans moved on to lesser gladii named for their manufacture in Mainz or Pompeii. But the *gladius hispaniensis* entered Roman folklore in the same way as previous legendary weapons, such as Gilgamesh's ax and the arms forged by the gods Hephaestus and Vulcan. The very existence of such weapons, to say nothing of their supernatural powers, suggested that the warriors who wielded them were doing God's—or the gods'—work.

The second most instructive and influential variation on the First Combined-Arms Paradigm of the classical age was the composite recurve bow, the weapon of choice for light cavalry. Like the chariot before it, this military instrument was invented by barbarians, probably on the Eurasian steppe. It was a short bow, with ends turned away from the archer, ideally suiting it to firing from horseback or chariot. The rider could easily move his weapon over the horse's neck or the chariot rail to fire from side to side. Like the *gladius hispaniensis*, it took its special qualities from the materials and the technique of manufacture. Its laminate construction—usually sinew in front, wood in the middle, and horn in back—maximized overall strength and power. These laminates were glued together, and then the whole bow was bent and steamed to impart its characteristic curve and wrapped to reinforce the structure. Short when unstrung and even shorter when strung, it was easy to carry and to maneuver when shooting. In the hands of experienced bowmen, it delivered enormous hitting power and high rates of accurate fire.

Thus it was that technologies within the First Combined-Arms Paradigm changed in significant ways while the paradigm itself stagnated. New variations on the basic arms and armor appeared and disappeared on the Eurasian battlefield, favoring now one combination of military force, now another. Through the rise and fall of classical Greece, the Macedonian Empire, and the Roman Republic, heavy infantry formed the center of gravity of Western armies. These phalangeal paladins, hoplites and their descendants, were collectively the queen of battle, the most powerful force on the chessboard of land warfare. Both Greeks and Romans encountered enemies in southwest Asia whose military formations were based on light or heavy cavalry, or both. Seleucids, Parthians, Armenians, Scythians, Sasanids, and others fielded swarms of lightly armored mounted warriors firing recurved bows, or troops of cataphracts (as the Greeks called them), armored warriors on heavy, sometimes armored horses, attacking with lance and shock.

Often the heavy horsemen were nobles or aristocrats, the members of society able to afford their combat panoply. In the heyday of the phalanx, disciplined heavy infantry had been proof against mounted attack. But, as the Western Roman military establishment deteriorated in the fourth and fifth centuries, the disciplined infantry formations of Rome's zenith gave way to more disordered European battlefields in which the mounted warrior rose in importance. The First Combined-Arms Paradigm abided—swords, spears, bows and arrows, shields, armor—but the center of gravity was changing sides. In the last centuries of the Roman Empire and first centuries of the Middle Ages, the infantry-cavalry cycle reversed itself once more. Mounted warriors became relatively more powerful, while infantry receded to supporting roles. As with the chariot, the imperative to change had come not so much from the technology of mounted warfare as from the waning discipline and training of infantry warfare.

Medieval warfare

In the fifth century CE, a dark age settled on Europe comparable to the dark age following the Catastrophe of 1200 BCE. Roman taxation and administration collapsed while authority, military force, economic networks, and political organization deteriorated. The First Combined-Arms Paradigm survived, but the transition from an infantry to a cavalry cycle slowed to a crawl. Three components of the new mounted weapon system emerged slowly between the fifth and the fourteenth centuries. First, the knight's armor transitioned in the late Middle Ages from the mail (garments woven from interlocking metal rings) of the late classical period into the plate armor of the Hundred Years War and finally full body armor for horse and rider in the sixteenth century. Second, the weight of this armor spawned horse breeding from the eleventh through the thirteenth centuries, leading to what historian R. H. C. Davis called "the age of the 'great horse'" in the fourteenth and fifteenth centuries. Targeted breeding in this period was accompanied by inclusion of more oats and other grains in the horses' diets, a change with logistic implications. While the lightly armed and armored mounted bowman of the Eurasian steppe could feed his smaller horse entirely on grass, giving the rider virtually unlimited mobility and range, the heavy mounted knight of the West was tethered more closely to his magazine and wagon train.

The third technical innovation behind the medieval mounted knight was the stirrup. This simple device, really a technological artifact, found its way from Asia to eastern Europe in the seventh century and western Europe in the eighth. One medieval historian, Lynn White, Jr., used its appearance in the West to complement a long-standing theory of the origins of feudalism in the West. German historian Heinrich Brunner had argued in 1887 that feudalism was in essence a social/political system based on a military relationship. The lord or king of a territory parceled out land to vassals (and they, perhaps, to subvassals) so that they could use the income from the property to pay for the expensive arms and equipment necessary to be a mounted knight. In return for this land and income, these vassals swore allegiance to their lord and promised forty days (or thereabouts) of military service each year. But why, asked critics of the theory, did European feudalism begin early in the eighth century? Because, White said, that was when the stirrup first appeared in the West, empowering the heavily armed and armored mounted knight to grow into the dominant force on the European battlefield. The stirrup allowed the knight to become a shock weapon, leaning into his lance and grounding infantry and mounted soldiers alike with crushing force. Lords gave mounted knights land, the revenue from which could pay for their expensive equipment and retinue, and the knights gave military service in return. The knight, in essence, was the nucleus of the feudal system, an ingenious and unparalleled concordance of political, military, economic, social, and judicial power. The stirrup made the mounted knight irresistible on the battlefield, and the mounted knight enforced the feudal system.

Critics of the Brunner/White thesis have dominated the literature for the last half century, noting that the mounted knight was already a dominant force before the stirrup, that the distribution of land to vassals was well under way when Charles Martel supposedly hit upon this formula after the battle of Poitiers (732 CE), that European feudalism was never the neat, uniform social system that the stirrup hypothesis envisioned, and that the saddle was more important for shock cavalry than the stirrup. Some scholars accused Lynn White of technological determinism, that is, of claiming that the stirrup produced feudalism. In fact, White explicitly rejected such a claim, proposing only that the stirrup, which seems to have appeared in the West at almost exactly the time of Poitiers, provided the final catalyst for medieval society to precipitate feudalism out of the political, military, economic, social, and judicial stew that was eighth-century Europe. The interpretive controversy over the impact of the stirrup offers a poignant reminder that "technological determinism" is usually a rhetorical flourish, never a historical reality. Historians accuse their colleagues of being technological determinists, but no respectable historian ever practices that interpretation. Instead, the judicious historian seeks understanding in context, drawing upon all those categories of analysis that promise explanatory power. The stirrup was a dual-use technology that helps explain—but did not cause—feudalism.

Historiographical controversies notwithstanding, there is no gainsaying that the heavily armed and armored mounted knight bestrode the European battlefield for half a millennium or more, from the early eighth century to the end of the twelfth century. The reason for his success was not so much the irresistibility of his military force as the psychological impact of his presence on the battlefield in the face of the motley, ill-equipped, and disorganized mobs that passed for infantry after Rome's collapse. In other words, this cavalry cycle appears to have imitated the chariot cycle that preceded it, enjoying dominance for half a millennium over infantry formations beset by disarray and fear.

The heavily armed and armored mounted knight began to experience serious reverses in the thirteenth and fourteenth centuries, before gunpowder finally dismounted him for good. First came the confederation of Mongol tribes that Genghis Khan (1162?–1227) molded into an imperial army. Genghis oversaw the conquest of northern China and central Eurasia to the Caspian Sea. His son and successor carried the Mongol conquest through Russia and the steppes all the way to modern Budapest, from which they threatened the land defended by the European feudal array. This Mongol army, barbarians all, simply outclassed the Europeans in every dimension of warfare. They had their own intelligence service; a sophisticated system of communication; a logistic train that supplemented their modest needs for human and horse food; an experienced light cavalry of mounted warriors who had spent their adult lives shooting animals and humans with composite recurve bows from horseback; a doctrine that blended dispersed strategic movement with tactical convergence to meet the enemy; a ruthless, bloodthirsty, and terrifying fighting ethic; and a leadership that traveled with the army and directed it brilliantly. What is more, the Mongols probably introduced Europeans to the most revolutionary of all military technologies: gunpowder. These invaders from the steppe brushed aside the West's mounted knights in Hungary and Poland in 1241 and seemed poised to extend the largest contiguous empire in all of human history from the Pacific to the Atlantic oceans. But then, in 1242, they suddenly turned around and retired to Mongolia. This civilization-saving reversal owed nothing to European resistance. Rather, the great khan had died, and all the tribes returned to the convocation that would choose his successor. Later Mongol attacks in Europe proved less effective, in part because of enhanced Western fortifications. Still, the Mongols of 1241 left the reputation of the European feudal array in shambles.

Nor did foreign invaders offer the only challenge to the European knight in the High Middle Ages. The Hundred Years War (1337–1453) pitted the English form of the feudal array against the

French form. The center of gravity in both systems remained the heavily armed and armored mounted knight. The English knights, however, enjoyed the support of a unique auxiliary: longbowmen. These lightly armed and armored foot soldiers wielded a bow of extraordinary proportions: about 6 to 7 feet in length at a time when the average Englishman stood perhaps 5 feet 6 inches. Their yew bows required prodigious strength and great skill to string, pull, and shoot with accuracy. But they could generate more than 100 pounds of force, enough to bring down any horse and penetrate all but steel breastplates. Furthermore, he could fire as fast as battle conditions required. To keep from being overwhelmed on the battlefield by swarming cavalry, the longbowmen of the fifteenth and sixteenth centuries sometimes planted stakes in front of their position, a barricade that could keep the enemy knights from running them down. Repeatedly, these bowmen proved to be the difference in battles with the French nobility in the Hundred Years War—most notably at Crecy (1346), Poitiers (1356), and Agincourt (1415). The French at times facilitated their own defeat by rushing the bowmen directly, without supporting missile fire and in tactical disarray. But it was the longbow that made the difference and allowed outnumbered English armies to roam the French countryside with impunity.

The continental feudal array met similar reversals when it tried to invade the provinces of Switzerland in the fourteenth and fifteenth centuries. There, militias met the mounted knights in highly disciplined and cohesive squares, bristling on all sides with pikes to arrest cavalry charges. Usually, the horses pulled up before impaling themselves on the pikes. All that was required was the courage and resolve of the soldiers to stand their ground. When the momentum of the cavalry charge was exhausted, soldiers armed with halberds, Lucerne hammers, morning stars, and other deadly polearms swarmed the milling cavalrymen. The hooks of their halberds could pull the knight from his mount. Once on the ground in plate armor, the knight could be dispatched easily with a knife through the eye-slit of his helmet or an ax through a vulnerable joint. Polearms with ax blades could cut through a leg of the knight's horse, bringing both to the ground and certain death or capture. In some battles, the feudal array outnumbered the Swiss pikemen enough to overcome them. But more often than not, the Swiss got the better of such encounters. When the mounted knight of the feudal array finally went out of fashion, many of the Swiss halberdiers, trading on their reputation, sold their services as mercenaries, most famously guarding the pope, as they still do today. While other royal bodyguards wear swords to evoke their historical lineage, the pope's Swiss guards still bear the murderous halberds that made them famous.

The European knight was thus defeated by three different counter technologies over the course of two hundred years before finally yielding pride of place once more to infantry during the sixteenth century. Why did this latest revolution of the infantry-cavalry cycle take so long to complete? Among the many answers are two directly related to the technology of European warfare in the High Middle Ages. First, the feudal system embodied a convergence of military, political, economic, cultural, and social power that gave it a robust institutional inertia. The mounted knight was the centerpiece of the system, with multiple levers of power at his disposal. Second, when the mounted knight could not dominate the battlefield, he could retire within his castle walls, to withhold feudal service to his lord, repel military challenges from his peers, and even withstand the Mongol invader. Medieval siege technology, hardly improved from its classical precursors, could overcome the fortifications of many towns and cities, but it failed more often than not against well-designed and well-defended castles. Furthermore, many feudal agreements limited the vassal's military service to forty days a year, hardly enough time to conduct the campaigns of starvation into which so many medieval sieges devolved. Perhaps the European knight of the High Middle Ages could not win all his battles, but he could always retreat to his inviolable castle.

The gunpowder revolution

Thus it was that gunpowder proved so devastating, not only to the individual knight but also to the whole feudal order. On the battlefield, individual firearms achieved more cheaply and routinely what the English longbow, the Swiss pikeman, and the Mongol mounted warrior did with their special weapon systems. And when the knight retreated to the safety of his castle, cannons blasted through the walls. These castles often had high, thin, curtain walls, testaments to the stagnation and ineffectiveness of siege technology since the time of the Assyrians. The walls were built high to deter ladders, but they were not built thick. When siege artillery rolled up before them, it easily opened gaps through which infantry could rush. Lords used artillery to bring their vassals to heel, to strip them of their exclusive hold on military power, and to convert their obligation for service in the feudal array into a tax. These lords could then use the tax revenue to raise their own infantry armies, buy more artillery, and subordinate more of the warrior nobility. Along the way, feudalism gave way to monarchy on its way to absolutism. Historian Clifford Rogers counts this as one of the most significant military revolutions in Western history.

This political and military transformation of Europe was but one of many changes wrought by gunpowder—the second of the three great military revolutions highlighted in this book, and one of the most important inventions of all time. At least eight other momentous consequences flowed from the gunpowder revolution. First, it inaugurated, in both warfare and society in general, an age of chemical power—what I call the Carbon Age. The cannon was the first internal combustion engine, powered, like most of its successors, by carbon-based fuels—wood (or charcoal, one of the ingredients of gunpowder) and fossil fuels (coal, oil, and natural gas). It revealed retrospectively that the technological ceiling capping the First Combined-Arms Paradigm was muscle—and to a lesser extent wind—power. In the future, the scale of warfare would expand to the limits of chemical power. Once weapons and other military technologies harnessed the power of chemical reactions—fire, for example—then a riot of death-dealing innovation ensued with world-changing speed. The scale of death and destruction unleashed by war through the remainder of the second millennium CE still beggars the human imagination. Though prehistoric warfare and warfare within the First Combined-Arms Paradigm killed more people per capita than warfare in the Second Combined-Arms Paradigm, most of those deaths had resulted from disease and famine brought on by war. Chemical energy unleashed on the world a killing power that climaxed in World War II with a "storm of steel" on the battlefields and the firebombings of Dresden and Tokyo in the urban centers of civilization.

Second, gunpowder changed the dynamic of fortification. Siege engines in the ancient and classical worlds had never been terribly efficient or effective. In Europe, at least, walls grew higher and thinner. Even the incomparable city walls of Constantinople gave way before the new firepower, contributing to the fall of the city in 1453. As the new guns became more powerful, the old walls became more vulnerable. Fortification had to change or fail. At the end of the Middle Ages, northern Italian city states pioneered a new form of fortification, the *trace italienne*, reigniting a contest of dueling technologies that had begun with the Neo-Assyrians and would continue into the twentieth century.

Third, missile weapons became deadlier than stabbing, cutting, or clubbing weapons. Though the bow and arrow had surely killed more humans than any weapon before the gun, it was still an instrument of pounce-and-flee tactics, disdained by the Greeks and associated with barbarians and the "light" auxiliaries of classical and medieval warfare. Now most of the killing on the world's battlefields would be done at a distance by firearms and artillery, a shift that appalled the likes of Miguel Cervantes—horribly wounded at the naval battle of Lepanto—and his fictional knight errant, Don Quixote. The strength and skill of the warrior might now succumb to the pull of a trigger finger. Courage and honor were forfeit to death at a distance.

Fourth, gunpowder dethroned the mounted knight while elevating the gunner. The cavalry of the Middle Ages gave way to a new infantry cycle. As Don Quixote feared, gunpowder put in the hands of any unwashed, unskilled commoner an instrument that could kill a noble knight. The shift in military power shook not just the warrior class but the whole society, putting commoners on top and the nobility at risk. Cavalry would not disappear until the twentieth century, but it would recede into the supporting role formerly occupied by the irritating but unmanly slingers, bowmen, and skirmishers of old. Not even the diabolical and deadly crossbow had had such an impact.

4. This painting depicts soldiers and contractors for the Polish army manhandling a primitive cannon over a pontoon bridge during the battle of Orsha (1514). Perhaps this was one of the weapons that surprised and turned back the Muscovite forces, handing the Polish-Lithuanian alliance an upset victory over superior numbers.

Fifth, a Second Combined-Arms Paradigm displaced the model that had dominated field warfare since the Catastrophe of the twelfth century BCE. The new paradigm added field artillery to the infantry and cavalry duo of old. From the seventeenth century until the end of World War II, commanders would juggle variations on three combat arms—infantry, cavalry, and artillery—all of them empowered by chemical energy and saturating the battlefield with firepower.

Sixth, the ammunition to fuel this firepower imposed on armies a logistical burden greater even than the demands of feeding heavy horses on campaign. One might find oats and other grains while traversing the countryside, but seldom would caches of arms, spare parts, fuel, and ammunition be found outside defended arsenals and magazines. Furthermore, the logistical tail following armies on campaign offered a vulnerability upon which a weaker enemy might pounce and flee.

Seventh, for the first time since the human community had divided itself into civilized, pastoral, and barbarian segments, the civilized states eliminated the existential threat posed by barbarians. Repeatedly in recorded history, supposedly primitive barbarian warriors had descended from the steppes of Eurasia or the deserts of North Africa to conquer great civilizations. Persians, Romans, Byzantines, Harappans, and Chinese had all succumbed at one time or another. Even Western civilization as a whole stood on the precipice of barbarian conquest in 1242. But never again. After the gunpowder revolution, barbarians might resist incursions by civilized states. They might even turn gunpowder weapons on their civilized enemies. But absent industrial know-how and infrastructure, they could never produce their own arms and ammunition. And lacking those, they could no longer threaten to conquer civilized states that had built up a gunpowder infrastructure. This asymmetry tempted many Western states into the imperial adventures accompanying the "rise of the West." Though many of those adventures ended badly for the Western imperialists, they never again faced extinction by barbarians at the gate.

Eighth, gunpowder transformed naval warfare with equally momentous consequences, as will be seen below. Ninth, and finally, gunpowder proved to be just the first phase of a larger, two-stage revolution of even greater impact. Gunpowder released the chemical power of carbon compounds explosively, sending projectiles flying at high speeds from the mouths of cannons, the barrels of small arms, and the shells of exploding devices. A second wave of carbon combustion would sweep over warfare in the nineteenth century, harnessing the chemical power of carbon compounds to drive machines of war. Those machines would achieve new heights of killing and destruction in the world wars of the twentieth century. That second revolution-within-a-revolution itself occurred in two phases, one in the nineteenth century and a second, more powerful phase in the twentieth century. That second half of the Carbon Age will be explored later.

But before turning to other realms of warfare, it is well to ask why it was that China, where gunpowder was invented, failed to develop its potential, while the West, which imported the concept, used it to such great effect. Historian William H. McNeill says that Westerners are simply a very warlike people. Kenneth Chase disagrees. Instead, he says, early firearms were too heavy and awkward to use effectively against the nomadic warriors of what he calls the "Arid Zone"—the Eurasian steppe and the North African desert. These are the people I call the barbarians. Instead, gunpowder technology favored those states facing the new infantry cycle—western Europe, Japan, and the Ottoman Empire. Thus, in his view, gunpowder fueled the infantry cycle and reacted to it. Robert O'Connell thinks artisanal entrepreneurs and capitalists gave the West its gunpowder edge. And there is no escaping the simultaneous origins of the scientific and gunpowder revolutions in the West, followed in many quarters by a riot of technological innovation. The West, after all, was the culture that came to view nature as something to be conquered.

Land warfare will be left at this juncture, at the beginning of the Second Combined-Arms Paradigm. The gunpowder revolution swept Europe in roughly century-long stages. Guns appeared in the fourteenth century. Siege guns toppled existing fortifications in the fifteenth century. Small arms sparked a new infantry cycle in the sixteenth century, dethroning the mounted knight. And mobile field artillery added a third combat arm to field warfare in the seventeenth century. A straight line led from that new paradigm to total warfare in the first half of the twentieth century.

Chapter 3
Naval, air, space, and modern warfare

Naval warfare

When naval warfare emerged from the mists of ancient history in the second millennium BCE, it was conducted in galleys, oared vessels fitted with auxiliary sails. As with warfare in the air and in space, and unlike most land warfare, naval warfare is defined by the platform carrying the naval warriors and their weapons. Naval warfare has been conducted on three classes of platforms defined by their systems of propulsion—galleys, sail, and steam—each with its own characteristic technologies and ways of warfare. One technology always governed the platform itself, the vehicle in which the warriors and their weapons rode to combat on an inhospitable sea, and one technology defined how the naval warriors fought. Their weapons might target the enemy ship or the enemy crew. But always the technology of the platform had to complement the technology of the fight. In land warfare, the chariot, and perhaps even the mounted warrior, might be seen as comparable marriages of platform and weapons. But before the twentieth century, the naval vessel was the most complex of military technologies—a system of systems.

Commerce called navies into being, both to attack it and protect it, and commerce has been a primary *casus belli navalis* ever since. Before there were naval vessels, civilian boats and ships carried people and cargo across the seas. The Mediterranean Sea, a laboratory of early naval warfare and an archive of its evolution, developed a particular kind of commercial ship, out of which naval vessels would evolve. These ships were frame-built; that is, the shell of the hull was built first, and then ribs and other stiffening infrastructure were added. This construction method worked in the comparatively calm seas of the Mediterranean, but it produced fragile ships reinforced with a keel external to the shell of the hull. The lightness and fragility made Mediterranean naval galleys fast and weak.

No doubt, piracy gave rise to naval vessels. Slow, unarmed merchant ships were vulnerable to faster raiding vessels that could overtake, grapple, and board a cargo-laden vessel. Pirates, in short, could pounce and flee. Adding soldiers to the merchant ships would have slowed them further and increased the cost of shipping without guaranteeing that they could stand up to maritime predators. Therefore, by the eighth or ninth centuries BCE, maritime states such as Assyria and Phoenicia were launching purpose-built naval vessels to protect their own commercial fleets and perhaps also to prey on the fleets of others. Soon these purpose-built vessels took on distinctive characteristics. For speed, they lengthened their hulls, decreased their draft, added more rowers, and converted the round bow of cargo ships into a ram. The tactic of choice was to ram an enemy vessel—civilian or naval—and then row backward to leave the victim holed and disabled. In time, the pointed and metal-sheathed prow—the Romans called it a *rostrum*—gave way to a blunt ram, meant to cave in the enemy hull instead of piercing it. Too often, the pointed ram could become wedged in the foundering hull of the enemy, disabling the attacking ship and allowing soldiers from the stricken ship to board and capture it.

As multiple states built up naval fleets in the classical era (500 BCE to 500 CE), an arms race settled on the Mediterranean. Speed was the main determinant of victory in this contest, and the structure of the Mediterranean galley dictated one way to achieve it: more rowers. Naval vessels grew longer, until they mounted fifty rowers to a vessel, twenty-five on each side. After that, the lengthening of vessels slowed, constrained perhaps by the scarcity of tall trees to serve as keels. Instead, rowers came to be stacked on multiple split levels in polyremes (Greek for "many oars"). Phoenicians and Assyrians floated biremes. The Athenians built the most perfect of all polyremes, the trireme, which brilliantly stacked three rowers in barely more than the fore-and-aft space normally taken by one. Thereafter, successor naval powers, such as Carthage and Rome, fought in quadriremes, quinquiremes, and even larger elaborations on the Athenian theme. Much doubt surrounds the arrangement of rowers on the monster vessels, but many scholars believe that the higher numbers simply meant they added more rowers per oar, rather than more oars per vessel.

5. *Olympias*, a reproduction of a Greek trireme built in the 1980s, enters the harbor of Tolon, Greece, in 1990. The galley was a weapon platform designed to ram enemy vessels with its underwater prow, but more often its crews disabled and boarded enemy warships for hand-to-hand fighting.

Whatever the propulsion scheme, these oared battleships surely exhibited the appeals and hazards of gigantism. If a weapon is effective, it seems, a bigger version will be more effective. This proved true for galleys—up to a point. The bigger galleys were built stronger to carry the additional weight of more rowers, possibly even big enough to mount siege engines for use against harbor fortifications. Their size also made them less vulnerable to holing by lighter galleys. And the larger polyremes also had more freeboard above the waterline, allowing their soldiers to fire missile weapons down on the main deck of enemy vessels and to jump down to board them. At some point, of course, the monster ships could be outmaneuvered by smaller, nimbler vessels, and the big ships could not chase pirates and other lesser craft into shallow water. Still, the galley battleship was a piece of monumental architecture, the most complex moving technological system of its day, useful for astonishing and intimidating those who might contest control of the sea.

As the ships evolved, their characteristics dictated tactical evolution as well. Ramming was always the ideal, but was seldom achieved. Alternatively, fast and nimble vessels rowed and steered by trained veterans might run the bow of their ship down the side of the enemy's vessel before he could ship his oars—that is, pull them inboard. The disabled warship could then be rammed, boarded, or bypassed while it tried to redistribute its unbroken oars. Often, however, even this tactic was beyond the capabilities of ships maneuvering in cramped quarters, and naval combat devolved into opposing fleets crashing into each other en masse in lines abreast. Then missile weapons and boarding parties would decide the issue. The Romans, who were primarily land warriors before coming into naval conflict with the Carthaginians in the First Punic War (264–261 BCE), found themselves at a disadvantage in both ships and seamanship. One mechanism they adapted from eastern Mediterranean naval warfare to cope with this asymmetry was the *corvus*, or beak, a pivoting gangplank that could project over the bow of their galleys. The device stood vertically on its pivot next to the forward mast, to be dropped onto the deck of an enemy vessel that came within range. Then Roman soldiers—the heart of Rome's military strength—could plunge across the gangway and take the foe in hand-to-hand combat. In short, they reduced naval warfare to land warfare on a floating platform—most famously at the critical battles of Mylae and Ecnomus. After building and losing three galley fleets, the Romans finally defeated the Carthaginians at sea and established what naval theorist Alfred Thayer Mahan would later call control of the sea.

Like other sea powers, however, the Romans discovered that control of the sea could be ruinously expensive. Galleys had a life expectancy of twenty to twenty-five years, and they consumed huge quantities of naval stores and rowers. Thalassocracies, dominant sea powers such as Athens and Carthage, might find the money to build and support standing navies—Athens devoted most of the revenue from a rich silver mine to pay for its fleet—but land powers such as Rome wearied of maintaining an army to secure its empire and a fleet to secure the Mediterranean, especially in the absence of a significant naval threat. Other states in subsequent centuries would struggle to find the proper balance between land and naval power.

After the fall of the Roman Empire, sea power in the Mediterranean fragmented among contending states and empires. The Byzantines came closer than anyone to controlling at least some parts of the Mediterranean. And they also defended their empire against challenges at sea with the only truly secret weapon of the ancient world. So-called "Greek fire" was an incendiary with many of the characteristics of modern napalm. As used on Byzantine *dromons*—small, fast galleys—it was preheated below decks and then shot under pressure from a nozzle in the bow. A flame at the tip of the nozzle ignited the fluid as it took flight. It reportedly stuck to anything it touched and continued to burn even underwater. Appearing in combat around 677 CE, during the first Muslim siege of Constantinople, it drove off the enemy ships and helped to secure the survival of the city and the empire. The formula for the incendiary was guarded jealously by the imperial family and their confidants, reserved exclusively for the defense of Constantinople until it disappeared, perhaps, in the riot of palace intrigues that passed for Byzantine government. No one then or since has been able to reproduce its reported effects. Even if those reports were exaggerated, enough people appear to have believed them to make this an unparalleled terror weapon.

Gigantism weighed upon galley warfare to the end. The Roman liburnian and the Byzantine *dromon* bucked the pattern, but the Venetian *gallia sotil* of the sixteenth century had about the same dimensions and twice the displacement of a Roman quinquereme. When the last great galley battle in Mediterranean history took place at Lepanto in 1571, the four participating navies tried to adapt cannons to their oared vessels. The Christian naval forces, averaging about five guns per ship, defeated the larger Muslim fleet, which averaged fewer than three guns per ship, but none of the participants found a convincing way to exploit the potential of shipboard cannons. The galley was going extinct in a new naval environment for which it was ill adapted. The gunpowder revolution that was transforming not just land warfare but all of world history had enormous potential to alter naval warfare as well. But it demanded a different platform. The biggest galleys at Lepanto could mount a large cannon facing forward on the centerline, but most of its gunpowder weapons were small antipersonnel guns. The casualties they could inflict seldom won battles. Cannon had the potential to kill ships, not just their crews. And in warfare on platforms, the platform is more important than the crew—John Paul Jones notwithstanding.

Two major and several minor technological innovations converged in the early modern era (1500–1800) to produce the Western side-gunned sailing ship, the most complex technological artifact of early modern history. First, the gunpowder revolution introduced a whole suite of weapons that worked as well at sea as on land. Second, the northern Atlantic sailing cog evolved in the late Middle Ages into a fighting platform, much as early rowed vessels had turned first into armed merchantmen and then into purpose-built naval vessels. The cog was a squat, rotund, slow, stable, seaworthy cargo vessel that had been carrying goods and some passengers about the Baltic and North Seas and up and down the Atlantic coast of Europe since the early Middle Ages. As this trade grew after the European commercial revolution of the fourteenth century, it attracted ever more piracy. The pirates attacked in similar vessels, carrying fighters armed with missile weapons and personal arms for boarding and hand-to-hand combat. The merchantmen naturally armed themselves in like manner, fueling a minor symmetrical arms race. To gain advantage in the resulting engagements, both sides built "castles" on the decks of their ships, structures from which their archers could fire down at the personnel on the decks of enemy ships. By the fifteenth century, individual firearms were being used from these castles, and it was not long before cannons were added to the ships' armaments.

But here the process hit a technological ceiling. Putting heavy cannon in the castles high above the waterline made the small ships unstable. And when the cannons fired, the recoil could tip the vessel precariously. So, only small antipersonnel weapons could be mounted thus. At some point a collateral innovation broke through this ceiling. Merchant shippers introduced ports in the sides of their vessels to facilitate loading and unloading of cargo. These ports were developed to be watertight when closed for sailing, to keep water from flooding in when the vessel heeled over under sail. Such ports, of course, could also be used for firing cannons. Once cannons could be moved from the castles above the main deck to the lower decks of the ship, then the firepower that a vessel might carry was limited only by the size of the vessel. A new race toward gigantism was on. From the hundred-ton cogs of the twelfth century mounting a few archers on primitive castles fore and aft there emerged by 1700 hundred-gun ships displacing almost two thousand tons. Cogs had targeted enemy crews. These floating batteries targeted other ships.

The full potential of the side-gunned sailing ship would never have been reached, however, were it not for a number of additional innovations. The steering oars that had maneuvered all galleys known to history gave way late in the twelfth century to a sternpost rudder, connected to a wheel or tiller on deck. This mechanism proved indispensable for controlling the large sailing vessels, whose handling characteristics in heavy weather put enormous stress on the helmsman. The compass had appeared in Europe around the turn of the thirteenth century. With it, sailors could venture farther from shore, eventually hazarding exploration of the Atlantic Ocean. In the eighteenth century the modern sextant replaced the more primitive astrolabe or cross-staff of previous centuries, giving mariners reliable estimates of latitude—distance north or south of the equator—when out of sight of land. And finally, also in the eighteenth century, Britain's John Harrison produced an exquisitely conceived and crafted maritime chronometer that could keep accurate time indefinitely, even on a rolling, pitching ship. With it, sailors could determine longitude, their position east or west of a fixed reference. Latitude and longitude gave the mariner an exact location in the middle of the ocean.

The capabilities of side-gunned sailing ships were unprecedented. Because they were powered by wind, a renewable energy source, their range was limited only by food and water for the crew. Since these were readily available worldwide, the sailing ship knew no bounds. As it roamed the world's oceans, it proved invulnerable to any other ship afloat, from Mediterranean galleys to Chinese junks to South Asian dhows. Indeed, so powerful were these ships that their only military competition on the high seas came from each other. An arms race of would-be European naval powers drove the size of vessels to staggering proportions, creating in the process a hierarchy of power. Fleets were dominated by "ships of the line," vessels of sixty guns or more that were big enough and powerful enough to survive duels with the largest vessels afloat. So expensive were these floating fortresses that only the wealthiest of states could afford to compete. Like the chariot, the ship of the line forced would-be naval powers to contend in kind, that is, symmetrically, or retire from the field.

The competition offered untold rewards and great hazard. Not only did sea power allow states to protect their own commerce and prey on the commerce of their enemies, as galleys had done, but it also allowed European navies to project power ashore. Thalassocracies such as the Dutch and the British prospered by concentrating their resources on naval power. States such as Spain and France tried to be great powers on both land and sea, like the Romans before them. They failed, suffering financial exhaustion and finally military ruin. At the end of the early modern competition for sea power and empire, Horatio Nelson, the greatest commander of the age of sail, defeated a combined French and Spanish fleet at the battle of Trafalgar in 1805. Though Nelson died of his wounds in the battle, Britain emerged as the unquestioned mistress of the seas, beginning a Pax Britannica that would last until World War I.

Nelson's flagship at that climactic battle of the age of sail was HMS *Victory*—a weapon system of one hundred guns, displacing 3,500 tons and requiring a crew of eight hundred to sail and fight. She descended directly from the North Sea cog that first put portals in her side to load and unload cargo. But *Victory* and her sister ships of the line existed under a technological ceiling that would constrain their operation and ultimately spell their doom. Like the galleys they succeeded, they were ruinously expensive. Building and maintaining them drove countries to denude their countrysides of trees and sweep their streets and taverns in search of destitute men who might be pressed into service as sailors and gunners. Their dependence on the wind for power limited their speed and direction of movement. They could go anywhere—but slowly, tacking back and forth to reach destinations upwind.

6. In the bittersweet, decisive battle of Trafalgar (1805), HMS *Victory*, the flagship of Horatio Nelson, was an icon for British command of the sea. The Irish artist Daniel Maclise depicted *Victory* as both the center of gravity for the battle and the idealized site of Nelson's death from sniper fire.

Furthermore, when they engaged in battle, they had to bring their guns to bear on the enemy by maneuvering the ship itself. In other words, their weapon platform was also their aiming mechanism. So they tended to fight in "line ahead" formation, with entire fleets sailing single-file parallel to the enemy fleet. Such tactics led to horrific exchanges of cannon fire, often at close range, often indecisively. Indeed, it was Nelson's willingness to attack the enemy line at right angles that brought about his greatest triumph.

By the time of Trafalgar, however, the age of sail was already in eclipse. The American artist/engineer Robert Fulton had built his first steamboat in Paris, and he would launch his first commercial model—the *Old North River*—less than two years after Nelson's death. Neither of these Fulton boats nor the others that sprang up in imitation and competition on the coastal and inland waters of the United States posed much threat to HMS *Victory*. The *North River* measured just 150 feet in length, and the weight of its chugging, vibrating engines threatened to shake it apart even in calm waters. The wooden paddle wheels driving all of these early steamboats through the water would disintegrate catastrophically under cannon fire. A battery of guns would stress them further, and high seas would sunder them before the enemy fired a shot. But they nonetheless began a technological evolution that produced within a century HMS *Dreadnought*, the first all-big-gun battleship. Just forty years after *Dreadnought*, the Japanese battleship *Musashi* turned turtle under fire and carried about a thousand of its 2,400-man crew to their deaths at the bottom of the Sibuyan Sea—the largest and most futile dinosaur in this particular race to gigantism.

The first phase of this story, from the *Old North River* to the *Dreadnought*, unfolded in the technologically fecund decades of the nineteenth century. The steamboat, a dual-use technology first developed for commercial purposes, proceeded on a single trajectory before specialized naval vessels were introduced. First the engines were made more powerful and more efficient by the introduction of double-acting pistons and high-pressure steam. The first steamboat made it across the Atlantic in 1819, but naval officers resisted the new technology as unreliable and aesthetically offensive. When steam entered naval ranks, it appeared mostly in the form of auxiliary engines mounted on traditional side-gunned sailing vessels of the line—a hybrid reminiscent of galleys with sails. Of course naval guns were improving at the same time, leading to experiments with iron armor mounted on traditional wooden hulls. In 1862, two armored vessels powered entirely by steam—one (the *Monitor*) purpose-built entirely of metal and mounting two guns in a rotating turret and the other (the *Merrimac*) an ironclad, side-gunned wooden vessel—met in the American Civil War to announce the arrival of the age of steam in naval warfare. Thereafter, developments proceeded rapidly: iron and then steel hulls, rifled guns mounted on centerline turrets, armored hulls and decks turning battleships into floating fortifications, turbine engines that converted steam into speeds of twenty knots or more, radios to communicate within and between fleets, gyroscopes to stabilize vessels and their guns, and range finders to allow one of these juggernauts of the sea to strike another 10 miles or more away.

Meanwhile, a new line of carbon-based heat engines transformed the nineteenth-century naval arms race and made possible the riot of power and destruction unleashed in the world wars. Nicolaus Otto pioneered the modern four-cycle, liquid-fuel, internal combustion engine in the 1860s, another dual-use technology. Now, for the first time, the stored power of carbon compounds—this time in the form of liquid fossil fuels—could be harnessed within the engine cylinder itself. Steamships would remain steamships, powered by water vaporized in a boiler immersed in a coal or oil fire. But the true internal combustion engine meant that much smaller machines could now be driven anywhere above the surface of the water and below the upper limits of the atmosphere with a compact engine running on liquid fuel. The applications included two technologies that would slay the mighty battleship.

One such application was the submarine. Attempts had been under way for centuries to build underwater weapons. Steamboat inventor Robert Fulton actually built one for Napoleon in 1800 and captained it and a crew of three into the English Channel in a failed attempt to blow up blockading British naval vessels. The Confederate States Ship *Hunley* succeeded in sinking a Union warship in the American Civil War, though it too sank in the operation. These ingenious devices were driven by human muscles turning cranks, hopelessly underpowered until internal combustion engines and storage batteries made them seaworthy and potent. American John Philip Holland pioneered the first modern prototype in 1897, just seventeen years before potent submarines began commerce raiding in World War I.

The internal combustion engine also powered the airplane, another dual-use technology invented for civilian purposes and quickly conscripted for military service. The Wright brothers taught the world to fly in a series of demonstrations in 1908. Within a few years, their invention raised human conflict into a third dimension and transformed the battlefields of World War I. In that same war, the tank harnessed the internal combustion engine to initiate a new cavalry cycle that would dominate land warfare through much of the twentieth century.

The internal combustion engine, and the machines it powered, transformed warfare more completely than its predecessor in the Carbon Age—the steam engine. Nowhere was the change more dramatic than in naval warfare. The steam battleship, like its predecessors of oars and sail, the polyreme and the ship of the line, swelled in its first century of existence from the diminutive, single-turret USS *Monitor* of Civil War fame to the unprecedented *Yamato* and her sister ship, *Musashi.* Both Japanese vessels displaced 72,800 long tons fully loaded—18 times the weight of Nelson's *Victory.* Their nine main guns, of more than 18-inch caliber, threw 3,200-pound shells 26 miles—100 times the weight and 26 times the range of Nelson's largest gun. Yet *Yamato* was sunk by carrier-based dive-bombers and torpedo planes that weighed about one-hundredth of one percent of their prey.

Gigantism had once again seduced warriors with a dinosaur. Repeatedly through human history, the allure of brute force had masked the potential of maneuver and aimed fire. In all realms of warfare, big and small were inherently neither good nor bad. Rather, technology magnified the brute force of big while enhancing the mobility of small. Technologies such as the internal combustion engine were great levelers and force multipliers. In the twentieth century and beyond, war became in many ways a contest to find the most appropriate technology for a given mission. Quality might or might not trump quantity. Big might or might not trump small. New might or might not trump old. The full arc of the Carbon Age was manifest in naval warfare, where change was sparked by technology push. That is, technologies developed elsewhere transformed naval warfare. Gunpowder had called into being the side-gunned sailing ship. The steam engine had called into being the steamship. And the internal combustion engine had powered the airplanes that sank the most monstrous of steamships.

So too did naval warfare figure prominently in the transition beyond the Carbon Age and into the Nuclear Age. Atomic or nuclear power was a dual-use technology of a rare sort: its inventors immediately appreciated its dual uses. As the atom gave up its secrets with accelerating speed in the 1920s and 1930s, physicists saw the possibility of destabilizing and splitting the atoms of certain heavy elements. If they could be split, the process would give off astronomical amounts of energy. If the neutrons released in the splitting of one atom could go on to split other atoms, the process might create a chain reaction. Theoretically, chain reactions could be controlled to release the energy slowly or explosively—one technique for power generation and one for bombs. When German scientists Otto Hahn and Fritz Strassman, with help from exiled colleague Lise Meitner, split an atom in 1938 by bombarding it with neutrons, scientists around the world quickly grasped the implications. With World War II looming in Europe, a race began to produce an atomic bomb. The resulting Manhattan Project in the United States, with collaboration from British and Canadian researchers, outpaced its rivals during the course of World War II and displayed its handiwork at Hiroshima and Nagasaki in 1945.

US Navy captain Hyman Rickover saw more clearly than most at the end of World War II that atomic power—what he came to call nuclear power—had military potential beyond the bomb. If the chain reaction could be harnessed, nuclear fission might power naval vessels, solving two problems that had shadowed the age of steam since its inception. First, a nuclear-powered ship could go years between refuelings, eliminating the need for frequent stops at coaling or oil stations. Second, a nuclear reactor could provide power without oxygen, making possible a true submarine, one that could remain beneath the waves for weeks, even months, on end. Rickover convinced the navy to let him find out if these possibilities could be realized.

After educating himself on the technology of nuclear reactors at the Manhattan Project's Oak Ridge National Laboratory, Rickover won approval for a pilot program to develop nuclear power for ships—beginning with a submarine power plant. The basics of the technology posed several pivotal questions from the outset. He would have to choose the fuel, a moderator to slow the speed of released neutrons, a coolant to maintain core temperature, a heat exchanger to transfer the energy from the pile to a water supply for steam to run the ship's turbines, a cladding material to reflect neutrons back into the pile and prevent the escape of radiation, and finally control rods to speed up or shut down the chain reaction. With the cramped quarters of a submarine in mind, Rickover chose a light water reactor. The pressurized water in this design performed cooling and moderating, while a different water supply provided heat transfer. For aircraft carriers, he would choose a similar but slightly larger boiling-water reactor.

It was the prototype light water reactor, built at Shippingport, Pennsylvania, that had the largest impact. Successfully developed for nuclear submarines, it powered the USS *Nautilus* on its maiden voyage in 1955. Thereafter, nuclear reactors powered two lines of American submarines, attack subs and so-called boomers, launching platforms for strategic ballistic missiles armed with nuclear warheads. When the missiles aboard these submarines scaled up to intercontinental range, the submarine-launched ballistic missile (SLBM) became the third and least vulnerable leg of the American strategic triad—bombers, land-based missiles, and SLBMs. Soviet weapons might strike American intercontinental bombers on the ground at their home bases or intercontinental ballistic missiles (ICBMs) on their launch pads, but never in the Cold War did the Soviet Union develop the capability to find and destroy the boomers hiding in the ocean's depths. The SLBM was the ultima ratio of Cold War deterrence.

Rickover's management of the Navy's nuclear reactor program had consequences beyond strategic deterrence. His choice of the light water reactor early in the program and the contract with Westinghouse Electric Corporation to build the first submarine reactors created a cascading effect. Economists often speak of this effect as "lock-in," a cultural or institutional commitment to one technological trajectory over another. Social scientists who study the history of technology call it "closure," by which they mean the end of a period of competing technological trajectories, when one is chosen and the others fade from prominence. Historian Thomas Hughes has called this phenomenon "technological momentum," to distinguish it from that hobgoblin of technology studies, the dreaded technological determinism. All of these analogies are meant to suggest that there is never anything "inevitable" about the technological choices communities make. Seldom is there one "best" technology for doing any particular job. Rather, different communities at different times and places find one technology better suited to their needs, resources, and temperament. But once those communities express a preference for one over the others, they create momentum behind that choice; they tend to close out further development of alternatives; and they lock the community into the investment they have made in their selection.

Thus it was with the light water reactor. Encouraged by the administration of President Dwight D. Eisenhower and its "atoms for peace" program, Westinghouse designed a light water reactor for commercial power applications. Many other combinations of fuels, moderators, coolants, cladding, control rods, and heat transfer mechanisms presented themselves, but Westinghouse capitalized on the technology it already knew. Thus it was that the United States launched its commercial nuclear power industry on a trajectory that helped determine its future. Economists call the resulting line of development path-dependent, meaning that the field is not free to find its own end point but is constrained by the path it started down early on. The further it goes down that path, the less likely it is to backtrack to the road not taken. First steps in path dependence are especially weighty, and Rickover made those fateful early commitments. For a multitude of reasons, America's love affair with nuclear power soured in the late 1970s, and the so-called second generation has yet to find much traction. One of the reasons was the choice made early on to suit military purposes.

Of course, the major factor behind the atrophy of the first generation of nuclear power in the United States was safety. The Three Mile Island accident of 1979 sounded the death knell of an industrial boom that was already struggling to maintain momentum. Rickover always insisted that nuclear power, properly managed, was safe. When he finally relinquished control of the Navy's nuclear power program in 1982, Rickover noted that, as he had promised, no US Navy ship was ever lost or even ever seriously damaged by a nuclear accident. The attack submarine USS *Thresher* sank in 1963, carrying her entire crew to the bottom of the Atlantic Ocean, but this was a mechanical failure not directly related to the ship's reactor. The navy's nuclear vessels avoided accidents through rigorous education, training, and discipline, overseen personally by Rickover. His whole career was a testament to human agency, to the power of people to limit the dangers seemingly inherent in certain kinds of technology. Modern complex technological systems can sometimes seem "autonomous" or "out of control," as political scientist Langdon Winner has suggested. But Rickover proved that humans could manage risk much more successfully than they typically do.

Air warfare

The airplane was a dual-use technology invented by two bicycle mechanics. It took the military a while to figure out what to do with it. Beginning in 1899, Wilbur and Orville Wright systematically scoured the existing literature on human flight. Then they designed and tested their own airfoils, developing lift tables to inform their design of both wings and propellers. To control themselves in flight, as birds did, they invented wing warping, an ingenious mechanism for achieving differential lift on their wings. From the ground they maneuvered a tethered version of their airframe in the wind until they understood control, then they mounted the big kite and glided in free flight from hilltops. By then, they knew how to fly. To power their airplane they commissioned a mechanic to design and build an engine to their specifications, and they designed their own propeller. They put all the components together in the winter of 1903 and flew a distance of 852 feet under their own power. Never before or since have two independent, untutored inventors read, thought, observed, theorized, experimented, and designed a new technology in such a short time with such staggering consequences. While their patent application awaited approval, they practiced flying for five years in a field near their bicycle shop. In 1908, they demonstrated their achievement in Paris and Washington, convincing all impartial observers that they had by themselves solved the problems that individual researchers, institutions, and governments had been attacking for years.

What did the world make of this gift of flight? Some used the airplane to look at the world from above. Civilians took pictures. Soldiers surveyed the battlefield. The U.S. Army Signal Corps, in charge of reconnaissance, purchased the first Wright Flyers. The possibility that this fragile platform of sticks and cloth could one day carry cargo and passengers, guns and bombs, beggared their imaginations.

Soon, however, researchers in Europe found themselves competing to launch faster and more maneuverable aircraft, part of a general arms race that accelerated Europe's descent into World War I. When new, fast airplanes rose above the battlefields of France, they encountered each other and quickly began fighting for what came to be called "air superiority." Observation planes morphed into fighter aircraft, mounting machine guns and reviving memories of chivalric knights engaged in one-on-one duels of honor. Since the death and destruction they inflicted in this air combat was limited to each other, the world took little note of the ominous implications for the future. The Germans experimented with bombing Britain from two huge, multiwing behemoths named *Gotha* and *Giant*. But the pilots over the battlefield dropped on the enemy troops little more than hand grenades and detritus from the floor of their open cockpits. The airplane of World War I was still primarily a vehicle, a platform for observing and driving off other observers. Only in the years between the world wars would this new platform find the military and civilian applications that eventually altered life on earth.

Two main military uses suggested themselves. Continental Europeans focused on fighter aircraft, driven by high-powered, liquid cooled in-line engines for speed and advanced aerodynamics for maneuverability. They sought to win control of the air over the battlefield to conduct reconnaissance and attack the enemy's ground forces. The United States and Britain, however, followed the lead of Italian air-power theorist Giulio Douhet, specializing in strategic bombing. This mission required an entirely different aerial platform, a larger vehicle powered by radial air-cooled engines to achieve great range. It was no coincidence that both these countries also needed commercial aircraft with the same capabilities: airliners to carry passengers across the United States and around the world to Britain's far-flung empire. The Germans, drawing on their experience in the Spanish Civil War (1936–1939), foresaw a greater scope for air power and developed a suite of aircraft for everything from air superiority, to long-range and medium-range bombing, to air assault by paratroopers. Eventually, the British had to complement their strategic bombers with fighters for home defense, and the United States had to add fighter escorts to protect their bombers on missions over enemy territory. Both

Britain and the United States experimented with fighter-attack aircraft for duty at sea. Unlike galleys and sailing ships, airplanes as platforms were all custom-designed from the outset for the particular weapons they were to carry and missions they were to perform.

An icon of dual-use aircraft between the world wars appeared in 1935, when the DC-3 took to the air. This passenger liner, the third iteration of a commercial design by the American Douglas Aircraft Company, featured cantilevered wings; cowled, air-cooled, rotary engines; variable-pitch propellers; retractable landing gear; wing flaps; a streamlined, stress-skinned monocoque fuselage; and flush riveting. It was the state of the art. Douglas built more than six hundred of the planes for commercial use before suspending production in 1942 to concentrate on military applications. It built more than ten thousand of the military derivatives—the C(argo)-47 and C-53—during the war and licensed the Soviets and the Japanese to build more than five thousand of their own national versions in the 1930s, aircraft which were converted to military purposes during the war. No airplane in history has matched the utility or longevity of the DC-3, which is still flying in some parts of the world.

The timeless qualities of the DC-3 contrast markedly with the hothouse evolution of purpose-built military aircraft. Air races and prizes, such as the one that lured Charles Lindbergh into his historic transatlantic flight in 1927, spurred technological innovation between the wars. Fighter, escort, reconnaissance, transport, and attack aircraft became more effective over both land and sea, as demand pull from the armed forces dragged technological capability into realms that airpower enthusiasts had barely imagined. During World War II, Americans moved up from the B-17 bomber with which they started the war—a plane capable of flying 287 miles per hour at 35,000 feet for a range of 2,000 miles with a bomb load of 6,000 pounds—to the incomparable B-29 of Hiroshima and Nagasaki, a behemoth flying 357 miles per hour at 32,000 feet for 3,250 miles with a bomb load of 20,000 pounds. But these marvels of military invention could not match all the predictions made by the early air enthusiasts. Strategic bombing never attained the decisiveness its prophets had predicted, and other forms of air power found their capabilities bent to practical purposes on the ground and on the sea. Close air support, for example, shaped land warfare more decisively than most had predicted, and air transport moved people and material around the world more quickly and safely than ships could do. Paratroopers jumping out of airplanes added a strategic mobility to ground warfare that no previous mounted warriors had ever achieved.

Additionally, suites of supporting technology arose to aid or counteract the rapid evolution of air power. Long-wave radar

7. This grainy action photo captures American B-29s dropping incendiary bombs over Yokohama, Japan, in May 1945. These "Superfortresses" could carry a bomb load of 20,000 pounds to a range of 3,250 miles and achieve a top speed of 350 miles per hour at 30,000 feet, above the altitudes that fighter aircraft could reach. Two of these planes, nicknamed *Enola Gay* and *Bockscar*, dropped the atomic bombs on Japan in August 1945. The B-29 was the ultimate weapon platform of its day.

along the United Kingdom's east coast gave critical early warning to British fighter-interceptors in the 1940 Battle of Britain. Shortwave radar, one of the most critical and fecund inventions of the war years, spawned more than one hundred applications, including airborne radar and proximity fuses. Improved radios allowed ground commanders to communicate directly with the close air support flying above them. New bombsights made possible what the Americans liked to call precision bombing—though it was never as precise in battle as on the test range. Long-range navigation systems (later called LORAN) guided bombers to distant targets. Antiaircraft weapons offered some ground defense against air attack. And, finally, atomic weapons debuted at the very end of the war, helping to spare the Allies and the Japanese alike the carnage of an amphibious invasion and giving the air power enthusiasts a patina of respectability for the frayed doctrine of decisiveness.

After World War II, radars merged with computers to form air defense systems that automated defensive responses while also promoting the dual-use technology of computer networking. Early experiments in pilotless aircraft laid the groundwork for unmanned aerial vehicles—sometimes called drones. Aerodynamic advances such as swept wings and wasp waists promoted supersonic flight. Guns and missiles vied for superiority in air combat. Aerial refueling extended the range of military aircraft.

Finally, aviation bequeathed to the world an institutional model of how to routinize and institutionalize innovations in military technology. Because flight posed more technological challenges than warfare on either land or sea, air power pioneered scheduled obsolescence on a cycle shorter than navies had yet anticipated. Even before one generation of airplanes was operational, its replacement was under development. Every new platform had to fly higher, faster, and farther than its predecessor; its weapon systems had to be more accurate, powerful, and irresistible; and its supporting technology had to improve safety, reliability, and efficiency. What one historian has called "capability greed" settled upon air forces sooner than other services and drove them into qualitative arms races of ruinous expense and intensity. And where they went, the other services followed, leading in short order to what President Dwight Eisenhower labeled a military-industrial complex.

Space warfare

The origins of space warfare paralleled aviation in many ways. Both were first invented by inspired amateurs as ends in themselves, without any attention to military applications. Soon, however, the military possibilities became apparent. Airplanes and spacecraft soon began serving as platforms for military activity, exemplars of dual-use technologies that found applications from both demand pull and technology push. The difference with spacecraft is that they never became the weapon platforms that early advocates of space warfare envisioned.

Spaceflight technology took off, so to speak, in the 1920s, spurred by theorists and visionaries who predicted that humans could and would travel to the moon, to Mars, and perhaps beyond. Reducing those visions to reality was the life's work of two remarkable communities: Wernher von Braun's Spaceflight Society in Germany and Robert Goddard's much smaller research team in the United States. Both experimented with early designs of small liquid-fuel rockets. Achieving some success, they sought outside funding for the more expensive proposition of building larger rockets that could carry significant payloads to very high altitudes. Von Braun and his colleagues turned to the Wehrmacht. Goddard solicited support from a wider variety of public and private patrons: the Smithsonian Institution, the United States Navy, and the Guggenheim family. The accelerating pace of military research and development in Germany and the United States in the years leading up to World War II carried rocket research toward weaponry rather than toward the civilian spaceflight that Goddard and von Braun had first intended. By the end of the war, von Braun's team had developed the V-1 (V for *Vergeltungswaffe*—vengeance weapon) "buzz bomb," a kind of cruise missile, and the even more famous and deadly V-2 ballistic missile, capable of carrying a 1,000-kilogram warhead about 200 miles. Germany launched more than three thousand V-2s at Allied targets in the closing years of World War II, but owing in part to the shortcomings of its guidance system, the Germans killed more of their own people in forced-labor camps building the rockets than they did bombarding enemies.

At the end of World War II, the United States and the Soviet Union captured most of the hardware and personnel from the V-2 program and put the people to work on their respective missile-development programs. The Soviets had greater need for long-range missiles, so they began an ambitious program in 1947 to build an ICBM capable of carrying an atomic bomb (which they had not yet perfected) from Soviet territory to the United States. The United States, in contrast, began its ICBM program in earnest only when it learned of Soviet progress. This particular race to develop symmetrical weapon systems was won by the Soviets, who displayed their achievement to the entire world on October 4, 1957, when they used their new ICBM to launch *Sputnik I*, a civilian, scientific satellite, into earth orbit. In spite of President Eisenhower's reluctance to militarize space, a hybrid military-civilian space race ensued.

The space race ran on parallel tracks, each relying on the dual-use technology of ICBMs. Configured as launch vehicles, they put the first satellites and then the first humans in space. At the same time, the military versions of these liquid-fuel rockets mounted nuclear warheads to join manned bombers as the second leg of a "triad" of the strategic weapon systems with which the United States and the Soviet Union fought their "Cold War" of deterrence from the 1950s through the 1980s. The third leg of the triad joined the suite in the 1960s, when the US Navy perfected a solid-fuel ballistic missile that could be carried safely aboard the

8. *Mercury-Atlas 6*, an American Atlas intercontinental ballistic missile configured as a space launch vehicle, lifts off from Cape Canaveral Air Force Station in Florida on February 20, 1962, carrying astronaut John Glenn in his *Friendship 7* space capsule on the first American orbital flight. The Atlas rocket is still in service, a dual-use technology supporting both military and civilian functions.

nuclear-powered submarines that Admiral Rickover had brought into being.

The United States won the civilian version of the space race in 1969, when it landed the first men on the moon. Those astronauts flew the Apollo launch vehicle, a civilian, purpose-built rocket masterminded by that chameleon of spaceflight, Wernher von Braun. He had been recruited by the National Aeronautics and Space Administration (NASA) to build launch vehicles for civilian, as opposed to military, masters. Thus, the cycle had come full circle, and the spaceflight enthusiast who wanted to go to the moon was freed from military service to pursue that goal.

The dual-use technology von Braun helped to pioneer had been advanced through World War II and the early Cold War by demand pull from the military, only to turn around in the 1950s and serve as a technology push for the US civilian space program. Turning his great talents and boundless ambition to civilian spaceflight, von Braun imprinted on the late twentieth century what one historian has called the von Braun paradigm. This model envisioned liquid-fueled rockets (descendants of his V-2) lifting people and material to low-earth orbit. There, astronauts would build space stations as bases for manned flights to the moon, Mars, and beyond. In the second decade of the twenty-first century, this model still guides the long-range planning of NASA.

Historian Walter McDougall has made clear that NASA and the civilian space program were really continuations of the Cold War by other means, but the United States at least tried to keep military and civilian space activities separate. Their Soviet counterparts organized their military space activities in the "rocket forces," creating a fourth branch of the military establishment on a nominally equal footing with the army, navy, and air force. The Soviet civilian space program was divided among a set of competing design bureaus, managed directly by the central government. This institutional arrangement in the Soviet Union contributed to the perception that space must surely become another arena of warfare—just as war had spread historically from land to sea and air, only in space, the platforms to carry military weapons would be more complicated, more expensive, and potentially more dangerous than anywhere on earth.

As it happened, however, warfare did not spill into space. President Eisenhower resisted domestic pressures arising from the "red scare" of the 1950s, the rampant alarmism of Joseph McCarthy and his fellow true believers, and dire warnings from the military-industrial complex that the militarization of space was inevitable. His immediate Democratic successors, John Kennedy and Lyndon Johnson, reversed the bombastic rhetoric of the 1960 presidential campaign to institutionalize the Eisenhower caution in a series of agreements and policies that slowed the American enthusiasm for arming the heavens.

By the time the Outer Space Treaty was signed in 1967, both superpowers and most other industrialized states in the world had agreed not to put weapons of mass destruction in space, not to make national claims on extraterrestrial bodies, and not to interfere with the orbital platforms of other states. The high cost of spaceflight, the vulnerability of spacecraft in orbit, and the difficulty of using orbiting spacecraft as platforms for earth-directed weapons had convinced virtually everyone that weapons in space were a bad idea.

Not until Ronald Reagan suggested a partially space-based antiballistic missile program in 1983—his Strategic Defense Initiative, quickly labeled by the press "Star Wars" after a contemporary science fiction film series—did a major power seriously consider placing strategic weapons in space. Twenty years later, President George W. Bush took that failed proposition one step further by withdrawing the United States from the Anti-Ballistic Missile (ABM) Treaty (1972), but by then the dismal cost-effectiveness of a space-based ABM system was already manifest.

Though weapons were not going to play a prominent role in space activity, earth orbit nonetheless became a site of critical military activity. Nonweapons technologies, ranging from reconnaissance satellites to global positioning systems (GPSs), grew in importance through the Cold War and into the twenty-first century. By the time of the Iraq and Afghanistan wars of the 2000s and 2010s, the United States military had come to believe that its dependence upon space-based assets for communication, intelligence, navigation, and weather monitoring had become so acute on the ground and at sea that the vulnerability of those assets posed a risk to American security. Through the turn of the twenty-first century, therefore, spacefaring states were developing new technologies to protect their assets in orbit and to threaten those of potential enemies. Space warfare may one day break out, but it is unlikely to appear until the technology of the von Braun paradigm is superseded.

Modern warfare

The American humorist Will Rogers observed, "You can't say civilization don't advance..., for in every war they kill you in a new way." This pithy insight resonates in the early twenty-first century as poignantly as it did between the world wars when Rogers voiced it. As this book has been at pains to emphasize, however, it was not true through most of human history. For thousands of years, the means and instruments of warfare evolved at a glacial pace. The gunpowder revolution in the western Middle Ages launched an epoch of accelerating change by adding chemical power to warfare. Modern warfare sped up the process and distributed it across four physical realms—land, sea, air, and space.

Historians conventionally divide the modern epoch into two phases. "Early modern" is usually associated with the period between the Middle Ages and the French Revolution, roughly 1500 to 1789. "Modern" is everything since, though some historians discern a postmodern period beginning sometime in the second half of the twentieth century. But postmodern warfare is not a concept with much explanatory power, so "modern warfare" will be treated here as a single epoch that is still running its course.

Some humanists and social scientists ascribe to modern history a set of characteristics they call "modernity." Among the most important features are Enlightenment rationality, secularization, the dominance of the nation-state, industrial capitalism, scientific and technological progress, and a particularly lethal and destructive form of warfare that targets both military and civilian sectors. In the twentieth century especially, the carnage of the world wars and the apocalyptic possibility of nuclear war cast upon "modernity" a shadow of ennui and foreboding, an apprehension that the human race might just disappear in a cataclysm of its own making.

In the nineteenth century, however, the benefits of modernity still seemed to greatly outweigh its hazards, especially in the Western states where "progress" was being made most rapidly. At its heart that progress was fundamentally material, an increasing understanding and mastery of the physical world that seemed to promise ever more wealth, comfort, safety, and health—at least for the Western states that were creating and defining modernity.

Philosopher of science Alfred North Whitehead said, "The greatest invention of the nineteenth century was the invention of the method of invention." He meant that the Western phenomenon of the scientific revolution followed by the industrial revolution had given rise to a technique of applying scientific method to technical invention. Problems were broken into their component parts, and each part was subjected to research in the existing literature, observation, hypothesis, experimentation, testing, innovation, and production. When all the pieces of the puzzle were in hand, they were integrated into a system of systems, and the whole was subjected to the same process. The Wright brothers' development of the airplane is a classic example. Of course this system of innovation worked as well for military technology as for civilian, but the Pax Britannica from 1815 to 1914 and the general absence of great-power war masked the murderous potential of the industrialization of warfare in the countries where it was taking shape. The colonized world and a handful of prescient observers saw what was coming, but most Westerners viewed the modernization of warfare as one more indicator of their superior civilization.

The transition in naval warfare from an age of sail to an age of steam, already discussed, offers something of a case study in nineteenth-century innovation. From the first successful steamboat—Robert Fulton's *Old North River* in 1807—to the British *Majestic*-class battleships of the 1890s, the guns, hulls, armor, propulsion, and size of warships had undergone a transition requiring an industrial, technological, financial, and administrative infrastructure that only the most developed nations could afford. Even mighty France essentially dropped out of the naval arms race at the turn of the twentieth century, leaving only Britain, Germany, Japan, and the United States as realistic contenders for sea power. Ambitious Russia thought that its antiquated fleet could compete; instead, it suffered at the battle of Tsushima in 1905 one of the most total naval defeats in all of history. Auxiliary technologies, such as automotive torpedoes, radios, gyroscopes, and high-pressure turbines further enhanced the war-fighting capabilities of the great navies.

Weapons driven by gunpowder, that most revolutionary of military technologies, also were experiencing dramatic change. On both land and sea, externally ignited, smoothbore, single-shot cannons and small arms gave way to self-actuating, repeating, rifled guns. In all cases, rifled barrels imparted greater range and accuracy than smoothbores, and prepackaged shells (or powder bags) with their own igniters allowed electrical or percussion detonation. In handguns, revolvers delivered multiple shots from pistols without reloading, followed before 1900 by magazine-fed pistols. Magazines did the same for rifles, either through human-powered reloading—bolt or lever actions—or through automatic gas or recoil mechanisms. Machine guns—fully automatic firearms—appeared by the end of the nineteenth century, ranging from the hand-cranked, proto-modern Gatling gun of the American Civil War to the Maxim gun, deployed to such great effect in the Boer War of 1899–1902. Americans, masters of laborsaving technologies, pioneered this mass production of death.

The technological transformation of ground warfare in the nineteenth century differed significantly from the change set in motion in the late Middle Ages by the introduction of gunpowder. Firepower was the key to victory in both eras, but increases in firepower resulted from different processes. In the late Middle Ages and early modern epoch, individual firearms and artillery were bulky, awkward devices, slow and difficult to reload in the face of the enemy. The barrel had to be swabbed clean after each discharge, loaded with powder, packed with wadding, and finally topped off by jamming a tight-fitting projectile down the barrel. Then a fuse or priming powder had to be ignited from an external source, vulnerable to wind and rain.

The key to firepower with such machines lay in training the gunners. Reloading small arms in the sixteenth century could require ninety steps or more. A unit of infantry hoping to achieve simultaneous volley fire could reload only as fast as the slowest gunner. Enemy cavalry could wait beyond the range of the guns and then cross the battlefield at a gallop to fall on the gunners before they were all reloaded. Infantry formations in the sixteenth century often posted pikemen in front of the gunners to protect them while reloading. The best-trained and choreographed gunners produced the highest rates of fire.

In the nineteenth century, increasing rates of fire flowed from the guns, not the gunners. Two hallmarks of modern technology—mechanization and automation—combined to saturate the battlefield with bullets. The soldiers actually were deskilled by the evolving technology. They needed only to aim and shoot, and volume of fire mattered more than accuracy. The Achilles heel limiting this torrent of projectiles was the logistics of providing ammunition to feed these shooting machines.

Chemists, mostly Europeans, introduced a variety of new propellants in the nineteenth century—the Americans called them smokeless powder—for both small arms and artillery. Gunpowder, or black powder, had always burned imperfectly, leaving solid residue to clog guns and thick smoke to blanket the battlefields of land warfare and the gun decks and turrets of naval warfare. Most of the new propellants pioneered in the nineteenth century were based on nitrated cellulose (gun cotton) combined with other ingredients to enhance stability and increase explosive power. The new propellants reduced smoke and residue and increased the power, range, and reliability of the guns. Standardized charges also allowed for the computation of more reliable firing tables, making it possible to site and range artillery more quickly and to "fire for effect" more efficiently. Predictably, these improvements in guns also spurred gigantism, both in siege weapons and naval guns. Indeed, the modern battleship taking shape at the end of the nineteenth century soon found itself in a race of dueling technologies—guns and armor—that would continue through World War II.

Non-weapons military technologies also contributed to the transformation of warfare in the nineteenth century. The steamboat, it must be remembered, began as a civilian, commercial technology, finding its way onto naval vessels slowly over the course of the century. More than half a century separated Robert Fulton's maiden voyage in 1807 from the first naval battle between steamships, the engagement of the *Monitor* and the *Merrimac* in Hampton Roads in 1862. The same pattern of civilian innovation was true of railroads, which originated hauling coal from mines. By the American Civil War they were moving troops and supplies within and between theaters of operation. Railway networks installed before the war favored the North, with lines running between eastern and western theaters, while the lines in the South, designed to connect the center with the periphery, provided avenues of assault for Northern invaders. In the wars of German unification, 1864–1871, a rail network intentionally designed to support military strategy sped Prussian forces and their gear to the frontiers and brought the wounded home.

Communication in the nineteenth century also served military and civilian purposes. The telegraph magnified command and control across and between theaters of operation, and furthermore empowered both military and civilian leaders to direct their subordinates in the field. The laying of submarine cables in the late nineteenth century extended the commander's reach internationally and also precluded tragedies such as the Battle of New Orleans in 1815, in which an estimated 336 British and American fighters were killed after the signing of the peace treaty in Ghent, Belgium, ending the War of 1812.

Countless other nonmilitary innovations arose in the fecund nineteenth century. Napoleon's army experimented with interchangeable parts, a technology that American inventor Eli Whitney claimed to have perfected in the first decade of the nineteenth century. In fact, Whitney's parts needed hand filing to make them truly interchangeable, but his example spurred successive innovators to perfect the technique. Humans had been manufacturing steel for more than three thousand years when a series of discoveries in metallurgy and innovations in manufacture made possible the mechanization and industrialization of made-to-order steel in large quantities and many forms. Captains of industry in the late nineteenth century—Krupp, Carnegie, and others—made fortunes manufacturing the steel for locomotives, skyscrapers, battleships, and artillery. Indeed, the industrial infrastructure of steel manufacture became a hallmark of both economic and military might. More mundane, but also transformative in its way, the quotidian tin can brought dietary variety and economy to people around the world and nutrition to soldiers on the march. The marvelous Montgolfier brothers of France even introduced the world of manned balloons, which quickly assumed military importance as reconnaissance platforms. Could the airplane be far behind?

The changes in warfare brought on by these rapidly evolving technologies had their greatest impact on imperial wars, where they gave Western powers an asymmetrical advantage over indigenous populations. Of course, the Western powers had enjoyed just such an advantage during the first wave of Western imperialism, beginning in the fifteenth century. The Spanish conquistador Hernando Cortés, for example, had projected Western power to the shores of Mexico on side-gunned sailing ships, and then conquered the entire Aztec Empire by capturing its capital with an invading army of a few hundred men equipped with gunpowder weapons, horses, and gunboats they built on site. In reality, Cortés owed as much to his indigenous allies as he did to his military technology, but the Aztecs were not the last natives to be awed and surprised by a gunpowder army.

Historian Daniel Headrick has noted that this first wave of Western imperialism put European powers in control of 35 percent of the world's land mass by 1800, the beginning of the modern era. Over the course of the long nineteenth century, those same Western powers used new technology to increase that control to 84 percent. The key to conquering half the earth's land mass in the course of one century, says Headrick, was power projection inland. Cortés's conquest of Mexico in the sixteenth century had been an exception. Most European conquest in the early modern era had been based on the side-gunned sailing ship. This instrument could sail into the ports of the major trading states in the nonindustrial world and seize control of the traffic in imports and exports. Appointment of viceroys, supported by an armed contingent with cannons and small arms, ensured that local rulers channeled wealth and commerce into and out of the ports in accordance with the mercantilist interests of the colonial power. The imperial power did not have to occupy the colonized state to make it serve the capitalist purpose.

Technological change, says Headrick, transformed this model in the nineteenth century. Steamboats allowed the Western colonizers to project naval power up the navigable rivers to the interior. The telegraph allowed the viceroys in the port cities and capitals to remain in touch with their inland outposts. The enhanced firepower of modern artillery and small arms ensured that small Western armies could prevail over large native forces armed with only the muscle-powered weapons of antiquity. Railroads carried those Western armies to inland entrepôts that the rivers did not reach. The Suez Canal shortened the lines of communication between the European states and their colonies in the east and south of Africa and Asia. Submarine cables put viceroys in touch with their home governments. And quinine—not really a technology—insulated the colonizers from some of the most debilitating indigenous diseases. In short, nineteenth-century technologies allowed the European imperialists to exercise control over entire territories and populations.

Technology's decisive impact on imperial wars in the nineteenth century proved less influential in great-power war. In part this was because the great powers—mostly European states plus the United States and, toward the end of the century, Japan—kept pace with technological change. The armies and navies of the industrialized states were armed symmetrically. They did not, in general, experience the asymmetry that dominated imperial wars. Even more importantly, there was simply not that much great-power war in the nineteenth century. The Pax Britannica that settled on the world between Napoleon's defeat at Waterloo in 1815 and the outbreak of World War I in 1914 reflected Great Britain's dominance of the world's oceans and Europe's exhaustion after a quarter century of the wars of the French Revolution and Napoleon. Two great exceptions to this pattern foreshadowed the ways in which technology would transform warfare between industrialized states, though many observers failed to appreciate just how complete the transformation would be.

The American Civil War (1861–1865) introduced many firsts. The North's industrial advantages over the agricultural South ranged from it superior transportation and communication networks to its manufacturing infrastructure, which could be converted to war production long before the Confederate states could build capacity from scratch. The North had a navy, along with the industry and shore establishment to support it; the South countered with innovative but inadequate experiments in blockade running, commerce raiding, mines, torpedoes, and submarines. The North countered or imitated these innovations, while adding armored riverine gunboats to its arsenal. Both sides revealed their entrepreneurial enthusiasm in the first year of the war by fielding steam-powered armored warships—the *Monitor* and the *Merrimac*—for a battle in Hampton Roads that is often taken as the transition point in the evolution from sail to steam navies. The North always had the overwhelming advantages of population and wealth, but these were amplified by its technological superiority, providing, in modern parlance, a "force multiplier."

The wars of German unification (1864–1871) provided another laboratory for changing military technology. In successive wars with Denmark, Austria, and France, Prussia shocked the world with the celerity and decisiveness of its victories. As with most great historical events, many causes lay behind the Prussian success—not least the militarization of the Prussian state, the poor preparation of its enemies, the professionalization of the Prussian army, and the ruthless geopolitical maneuvering of Otto von Bismarck (1815–1898), the minister president of Prussia, who isolated rivals and blocked outside interference. Technology played an operational role, especially in the strategic use of railroads and the high quality of the Prussian Dreyse needle gun, which allowed soldiers to reload in a prone position. By the time of the Franco-Prussian War (1870–1871), however, another emergent feature of modern industrialized warfare trumped this Prussian advantage—the clash of dueling technological developments. The French chassepot rifle proved just as effective as the Dreyse, pitting symmetrical small arms against each other and nullifying, to some extent, the advantage that each side had hoped to achieve with its innovation.

Most observers marveled at the power on display in these midcentury great-power wars in America and Europe. They fit neatly into a larger narrative of human dominion over the forces of nature and the peoples of the undeveloped world. In the closing decades of the nineteenth century, Westerners treated themselves to dozens of international fairs celebrating their technological prowess. The London exposition of 1851 established the model, and imitators proliferated as the century wore on. In historian Michael Adas's phrase, Europeans came to see "machines as the measure of men," proof of the superiority of their civilization and also justification to take over and make over the rest of the world in their image.

Late in the nineteenth century, concern about the growing lethality of modern weaponry sparked an interest in arms control that would continue through the twentieth century and into the twenty-first. Since the beginning of history, and perhaps before, human societies had agreed to limit their use of some military technologies and techniques when fighting against others they considered to be like them. Usually, however, war against strangers or barbarians or "the other" was without constraint. The famous Christian sanction of the crossbow at the Second Lateran Council in 1139, for example, condemned the use of the weapon against other Christians while allowing it against Muslims. The Hague Conventions of 1899 and 1907 limited the use of poison gases, bullets that expanded on impact, projectiles dropped from balloons, arming of merchant ships, and laying of automatic contact submarine mines. The effectiveness of such prohibitions has always been limited by the assertion that the laws of war may be overridden by "military necessity." Arms controls have always worked best when the participating states have seen it as in their best interest to comply.

At the turn of the twentieth century, however, only the most astute monitors of contemporary history saw clearly where the evolution of military technology might lead. The best of these, Jan Bloch (1836–1902), a Polish banker and financier, predicted in a multivolume analysis of nineteenth-century warfare that war was bound to lose its decisiveness. Escalating firepower, both small arms and artillery, would drive combatants to ground and eliminate maneuver on the battlefield. Armies would grow in size and power. The field of battle would expand beyond the ability of the commander to see or comprehend. And industry would feed the huge armies with endless supplies of food and ammunition. The result, said Bloch, was bound to be static wars of attrition, ending not in victory but in mutual exhaustion, both moral and economic. At the heart of the stalemate was technology.

Total warfare

Bloch's ominous prophecy unfolded much as he had anticipated in the trenches of the Western Front in World War I (1914–1918). There the combatants of the great powers saturated the battlefield with shrapnel and bullets, driving the soldiers underground in lines of excavation that zigzagged from Switzerland to the sea. Both sides tried innovations in strategy, tactics, techniques, and technologies to break the standoff: artillery barrages, chemical weapons, commerce warfare at sea, strategic bombing, strategic misdirection at Gallipoli, pioneering tactics of fire and movement, and even primitive tanks. Nothing worked. Just as Bloch had predicted, moral and economic attrition finally determined the outcome.

World War II (1939–1945), sometimes viewed as the second phase of a single great-power world war, proved even more titanic and transformative, a watershed in human history. To begin with, both conflicts were humanity's first and only total wars. Second, they were wars of industrial production, won by the alliances that produced the most stuff. Third, World War II was the first war conducted in all four realms of human warfare: land, sea, air, and space. Fourth, World War II was the first war in human history in which the weapons in play at war's end differed significantly from those at the outset. Fifth, World War II was the last great-power war. And sixth, World War II ended with the nuclear revolution, a turning point in the technology of warfare and the history of humankind.

Journalist/historian Walter Millis (1899–1968) argued that "total war" is best thought of as a culmination of three great historic revolutions. The French Revolution had introduced the *levée en masse*, the nation in arms. The industrial revolution had shown how to produce enough war materials to arm, equip, and move those mass armies; Fordism and mass production of the twentieth century had only improved upon the speed and efficiency of industrialization. And the Prussian general staff had introduced a managerial revolution equal to the task of marshaling those forces in time and space. Only when these capabilities were in place would five thousand years of evolving military technology climax in the carnage and destruction of the world wars.

Both of the world wars were contests of industrial production, a convergence of modernity with industrialization and twentieth-century mass production. Along the way, each side tried to use its vast arsenal to destroy not just the arsenal of the enemy but also the enemy's will and material resources. Inescapably, the enemy's population became a target, for it embodied both the national will and the productive capacity of the state. Never before had so much material been mobilized for warfare, and never before had so much been destroyed. Most wars in human history have killed more by disease, famine, and dislocation than by direct assault, but the world wars were different. Killing and destruction had themselves been industrialized, laying waste combatant states on a scale never before seen. The Axis powers finally ran out of stuff before they exhausted the will of their populations.

For most of history, humans fought only on land. Not until the late ancient or early classical periods did interstate conflict go to sea. Another two millennia of technological development passed before humans flew for the first time, but it took hardly more than a decade to pass from the first flight to the first air warfare. Space warfare followed in just a few decades: the V-2 rocket could and did fly into space, though its wartime trajectories kept it within earth's atmosphere. The world wars introduced two new realms of warfare in just half a century, fueling human foreboding that technology was out of control. Some students of modern military technology classify cyber warfare as yet a fifth realm, but it is as old as electromagnetic control mechanisms, which were also at work in World War II.

World War II witnessed the development and introduction of significant new weapons that did not exist before 1939. The list includes microwave radar, jet propulsion, proximity fuses, guided missiles, cruise missiles, "precision" bomb sights, acoustic torpedoes, computerized code breaking, and, of course, the atomic bomb. The important point to note here is the appearance of systematic, institutionalized military research and development in World War II.

World War II was the last great-power war. Indeed, there has been very little interstate war of any kind since 1945. Most war has been intrastate: rebellions, insurrections, civil wars, and anarchy within failed states. Seldom in these conflicts are the total resources of the state mobilized, as they were in total war. The technology of this kind of warfare is generally "conventional," sometimes ad hoc. That is, it deploys the same Combined-Arms Paradigm as in World War II: infantry, artillery, and mounted warfare (tanks, personnel carriers, and later helicopters) supplemented by tactical rockets and missiles and close air support. The absence of great-power warfare since 1945 has contributed to this new technological stasis, as has the most revolutionary of modern military technologies: nuclear weapons. Appearing at the very end of World War II, nuclear weapons ushered in a "long peace" (John Lewis Gaddis's name for the absence of great power war) that abides in the second decade of the twenty-first century. Many factors contributed to this long peace: the destructiveness of conventional, industrialized warfare, the creation of new international institutions such as the United Nations, a growing commitment to the rule of law, the increasing interconnectedness of the community of nations, acceleration of communication and transportation, and growing appreciation that modern war was no longer winnable.

But none of these factors had the clarity, immediacy, and materiality of Hiroshima and Nagasaki. As the Cold War incited an arms race in the two decades after World War II, as the superpowers moved from atomic bombs to thermonuclear bombs—an order of magnitude lighter, cheaper, and more powerful—and the weapons proliferated to other states, a taboo against nuclear war settled on the human community. Mankind had finally developed a weapon too horrendous to use. If the illusion of winning modern war had not been clear before the world wars, Hiroshima and Nagasaki revealed it starkly. Even as the two superpowers amassed a combined nuclear arsenal of seventy thousand warheads, a consensus coalesced around never using them again.

Seventy years into the nuclear age, the consensus holds. Nuclear weapons still exist, and they proliferate slowly. But they have served so far as guarantors of interstate peace. There have been no great-power wars and only a handful of interstate wars in the last seven decades, wars that have been limited by international cooperation. Of course, the long peace could end at any time, and the nuclear taboo could fail. Ideologues bent on suicidal terrorism might one day acquire an atomic or thermonuclear device—or some other weapon of mass destruction. But if they ever detonate such a weapon in anger, they will no doubt discover that nuclear weapons offer deterrence and retaliation, but they serve no useful offensive purpose. For now, at least, the nuclear revolution has produced a more peaceful age than any mankind has ever known, the daily carnage on the evening news notwithstanding.

Chapter 4
Technological change

Research and development

In addition to generating new weapons, World War II ushered in two momentous transformations in the world's relationship with military technology. The nuclear revolution, already mentioned, will be addressed again later. The second great transformation of military technology was modern, institutionalized, routinized research and development.

World War I saw some mobilization of scientific and technical research, but it was nonetheless a war of industrial production. In many ways, World War II followed the same pattern, with the United States serving as the great arsenal of democracy. Its gross domestic product (GDP) exceeded the combined output of all its major allies—Britain, France, and the Soviet Union—plus all the other states that comprised the United Nations. By the end of the war, those states had a collective GDP five times the size of the collective output of the Axis powers. The war in the North Atlantic finally shifted in favor of the Allies in the first half of 1943, when they produced ships and cargo faster than German submarines could sink them. The Germans lost the pivotal Battle of the Bulge in the winter of 1944–1945 when they had to abandon their tanks on the battlefield for want of fuel, while the Allies had already run a fuel pipe across the bottom of the English Channel to power the armada of tanks and trucks that would motor to Berlin. Napoleon's army may have traveled on its stomach, but military forces in the middle of the twentieth century traveled in ships, planes, and motor vehicles powered by internal combustion engines and fueled by petroleum. Their logistical tail stretched behind them in a never-ending umbilical back to Roosevelt's arsenal.

This multiheaded juggernaut, powered by Allied—especially American—industrial capacity, did not, however, always bring to bear the best war materiel. American naval aircraft and torpedoes, for example, were inferior to those of the Japanese. American tanks were inferior to both German and Soviet tanks. Both Germany and Great Britain flew jet aircraft before the end of the war; the United States did not. German long-range submarines were every bit as good as American ones. And late in the war, as Germany collapsed under the weight of Allied stuff, Hitler channeled his dwindling resources into secret weapons, new technologies that might yet turn the tide. His jet aircraft, especially the ME-262, had the potential to deny the Allies air superiority over the battlefield, but there were too many bugs, too little fuel, and not enough airplanes for them to pose a serious threat. Wernher von Braun's rockets could reach targets in Britain, but they were not sufficiently accurate or numerous to cow the British. Still, the potential of these new weapons demonstrated that the Allies did not have a monopoly on military invention and innovation. When the atomic bombs exploded over Hiroshima and Nagasaki to end the war in the Pacific, they confirmed the decisive impact of research and development on warfare.

The United States came away from this experience chastened by the many instances in which its military technology proved inferior to the enemy's. For all the remarkable achievements of its scientific and technical establishment during the war, including the incomparable Manhattan Project, military leaders nonetheless concluded that they could not return to the prewar mechanisms for developing new military technologies. Quantity had been the main determinant of victory in the world wars, but quality could determine the outcome of the next war.

The leader of America's wartime mobilization of science and technology was Vannevar Bush, head of the wartime Office of Scientific Research and Development and de facto science advisor to President Franklin Roosevelt. At the end of World War II, he wrote for the president *Science: The Endless Frontier* (1945), a blueprint for government support of American research and development for military, medical, and economic innovation. Bush's experience during World War II had convinced him that scientists knew best. The government should fund a "National Research Establishment" and let the scientists set the agenda. The American government, unwilling to give any group such carte blanche, rejected his plan, opting instead for a National Science Foundation (NSF) and the National Institutes of Health (NIH) for basic research in science and medicine, respectively. Most other government-funded research and development was left to the mission agencies, such as the newly formed Department of Defense. The NSF and NIH would do "basic research" in pursuit of general understanding—a kind of technology push—while the mission agencies would apply demand pull to bend the potentials of science and technology to their specific needs.

Within the Department of Defense, each of the three main services—army, navy, and the newly independent air force—quickly developed their own idiosyncratic mechanisms for promoting technological innovation suited to their institutional goals and doctrines. The army, the least technological of the services, elected to continue its support of wartime contractors, such as the Moore School at the University of Pennsylvania, working on computer development. It also created some internal infrastructure—first a Research and Development Division and then an Office of the Chief of Research and Development—to oversee its activities. But otherwise it continued to rely on its time-honored arsenal system to provide innovation.

The navy proved to be the most progressive of the three services, exploiting the ties it had long since established with universities and other institutions of basic research. It continued and enlarged its wartime Office of Naval Research and expanded the operations of its fabled Naval Research Laboratory in Washington, DC. These complemented the navy's established programs of ship design and development, which evolved into the Naval Sea Systems Command.

The air force, successor to the wartime Army Air Forces, took the most dramatic steps toward a new model of government-supported innovation. Continuing the wartime pattern of relying on contracts, it first recruited Theodore von Kármán, the legendary aerodynamicist at the California Institute of Technology, to chair a scientific advisory board and produce a twelve-volume study of the future of aviation. The title of volume 1, written by von Kármán, said it all: *Science: The Key to Air Supremacy.* The air force went on to buy innovation by contract, even sponsoring the RAND (Research and Development) Corporation, the first of the think tanks that would become mainstays of US military research and development. The air force also continued its army tradition of arsenals, expanding its in-house research and development (R&D) program at Wright Field in Dayton, Ohio, and opening new laboratories such as the Arnold Engineering Development Center in Tullahoma, Tennessee.

So great was the enthusiasm of the military services for new technology that the secretary of defense felt compelled to place institutional constraints on reckless innovation. A Research and Development Board, called for by the National Security Act of 1947, which created both the air force and the new Office of the Secretary of Defense, evolved in the Eisenhower administration to become an assistant secretary of defense for research and development. This office, under varying names, has been in existence since 1953. After *Sputnik I* and the riot of space proposals made by the different services, President Eisenhower concluded that another agency was required just to sort out the competing schemes for technological one-upmanship. The Advanced Research Projects Agency (ARPA) came into existence in 1958 to screen the half-baked proposals coming from the services, but it too developed a life of its own and took on the additional role of seeking out and promoting new technologies deemed to be in the national interest.

With this sort of institutional promotion, defense research and development grew like Topsy during the Cold War, producing some monstrosities of technological excess. After *Sputnik I*, the army proposed building a base on the moon, because it was a staple of military theory always to take the "high ground." The army and air force in the 1950s found themselves embroiled in the so-called Thor-Jupiter controversy, each one spending lavishly to build its own version of the same intermediate-range ballistic missile. Interservice rivalry over defense dollars spurred technological innovation as a bureaucratic technique for capturing roles and missions and the budgets that went with them. The marine corps insisted on building vertical-takeoff-and-landing aircraft—the Harrier and the Osprey—that proved to be more expensive than useful. The navy insisted on pursuing a nuclear-powered fleet even though this form of propulsion proved too costly for most ships other than submarines and perhaps aircraft carriers. The air force tried to develop a piloted, reusable space plane; unfortunately, they chose to name it "Dynasoar," for "dynamic soaring," and a dinosaur it was. The air force's twenty-first-century extravagance, the F-35 multipurpose fighter, threatens to bankrupt the service.

This hothouse environment of technological enthusiasm and development was labeled by President Eisenhower in his 1961 farewell address the "military-industrial complex." By this he meant that defense contractors and the military services had fallen into a liaison built around their shared interest in exaggerating the dangers of the Cold War and promising security through expensive, cutting-edge technology. Many observers since have noted that Congress and America's universities were also complicit in this "complex." Members of Congress found it useful to promote military R&D and production in their states or districts, and universities found it useful to accept research funding from the military services. Nor was the United States unique in this regard. One historian has seen in Britain during the Cold War a "warfare state," and a Berkeley political scientist noted in the depths of the Cold War that the United States might *have* a military-industrial complex, but the Soviet Union *was* a military-industrial complex.

In short, the competition for new and winning military technologies drove the great powers to institutionalize military innovation. At times this entailed beginning work on a next-generation weapon system as soon as the new generation went operational. This planned obsolescence mirrors the annual design changes promoted by the American automobile industry in the 1950s and 1960s. "Capability greed" became a staple of weapon systems specifications, driving costs up and reliability down. The military establishments of the industrialized states found themselves racing not so much against each other as against themselves. Some incentive was provided by the international arms market, which had a seemingly insatiable appetite for the newest military technology. Most of the incentive, however, bubbled up within each state's self-reinforcing military-industrial complex.

The passage of time resolved some of these disputes. The Cold War ran its course without triggering the third world war that many had feared. Indeed, the world backed into John Lewis Gaddis's 'long peace.' In war or peace, however, the military captured most US government spending on research and development. This pattern is controversial on many counts. Costly development of questionable technologies starves basic research that might produce more long-term, fundamental innovation. Many economists feel that military R&D tends to produce less economic growth than investments in civilian realms such as energy, transportation, and infrastructure. Civilian R&D is more likely to spin off military applications than vice versa. And military R&D tends to be gold-plated, because the contracting parties are operating in a marketplace that is simultaneously monopsonistic and oligopolistic: there is one buyer given to "capability greed" and a small number of sellers given to nonprice competition.

Dual-use technologies

In addition to having both military and civilian applications, some technologies are dual-use in another sense: the military may use them in both weapons and non-weapons roles. Some examples already have appeared in this book: fortifications, roads, chariots, steam engines, transport aircraft, and nuclear power—to name a few. But this category of military technology warrants closer examination, for it places the technology of warfare in the larger context of technology in general, and it illuminates one important dimension of the timeless dialectic between the military and civil society. Just as societies get the armies they deserve, they also get the military technology they deserve. Furthermore, many civilian technologies arise from military sources, shaping civil societies in ways seldom explored.

Non-weapons dual-use technologies

Non-weapons military technologies are the most obvious candidates for dual-use. They support warfare without attacking people or things. One inexorable trend in warfare has been the growth in number and significance of non-weapons military technologies. The earliest warfare no doubt began with the simplest of instruments—spears, knives, clubs, stones, and bows and arrows—all weapons. Very little support was needed. Over time, however, communities found that warriors were more successful when supported with armor, logistics, intelligence, communication, medical treatment, transportation, and the like.

As these services and supplies multiplied, warriors came to be seen as the "tip of the spear." In time the shaft came to outweigh the point until, in the twenty-first century, support personnel and material can account for 90 percent or more of a military force. In modern parlance, this is called the tooth-to-tail ratio—the balance between fighters and enablers. Though military culture continues to extol the primacy of the warrior who delivers the kinetic impact to the target, the truth is that non-weapons technologies outnumber and outweigh those at the tip of the spear. A few examples will make the point.

Fortifications, the most influential of non-weapons technologies, have already appeared in this story. They did not so much determine who won or lost a war—though they sometimes had that effect—as they determined when some wars would happen and, more importantly, not happen. As states and civilizations separated themselves from the barbarians and pastoralists who continued to live beyond the pale, they built cities dominated by monumental architecture—including walls. The walls, of course, were artifacts, not technologies. But they shared with the temples and ziggurats and public forums building technologies of large structures. Whether made with stones (Jericho) or dried bricks (Uruk) or concrete (Rome), these cities almost always had walls to hold the barbarian at bay. The walls, like those that protected Constantinople for more than 1,100 years, were designed both to repel and to intimidate would-be attackers. They announced that the residents of the city had power and resources equal to any challenge. They were, in short, a deterrent to war, a promise of futility and defeat to any who dared assail them. And as civilizations grew in military power and sought to conquer each other, they strengthened their walls all the more, to send the same message to their peer states.

Some states chose to fortify not only their cities but also vulnerable portions of their borders. The Great Wall of China, erected over the course of a thousand years or more, extended in various overlapping segments more than 13,000 miles along the northwest frontier of China. The Romans erected their own border walls, called *limites*, across the natural boundaries and invasion routes into the empire. Originally roads punctuated by defensive watchtowers and forts, the *limites* were sometimes elaborated with palisades and occasionally stone or earthen walls, like Hadrian's Wall in Britain. Like the Chinese, the Romans did not expect to stop invaders so much as deflect and slow them down, so that armies could be dispatched to the frontier to confront them. This is not so different from the purpose of the infamous Maginot Line, built by France between the world wars. Though the Maginot Line gave static defenses a bad name when it was circumvented by invading German armies in 1940, it actually did what it was meant to do—slowed and deflected the invader until reinforcements could arrive. Unfortunately for the French, their army still could not hold.

Fortifications in history have given the civilizations that built them an added benefit. They have allowed their states to reduce the standing armies that they would otherwise have needed to defend themselves. The walls served, in short, as a peacetime investment in security that paid dividends in all the years beyond their period of construction. States capable of extracting sufficient revenue or labor from their citizens could build defensive public works while holding down the much higher cost of keeping expensive soldiers under arms in the absence of a threat. And because fortifications had little offensive power, they were an investment in peace—that is, in military technology that did not directly threaten their neighbors.

Perhaps the second oldest and most important non-weapons military technology is roads. Like fortifications, roads are not technologies but artifacts of technology. Indeed, the first roads were not even that, but simply the routes, like the Silk Road, that humans and animals traversed with enough regularity to leave a trace. In time, civilizations began improving these natural thoroughfares. When those improvements reached the level of using tools and machines in a conscious technique to produce a solid and durable roadbed, then a technology of roads was in place. Archeological evidence of such roads comes down to us from Persia, China, Peru, and other empires. The Romans raised this technology to high art, disposing the same practical engineering that marked the Colosseum, the aqueducts, and the purely military technologies of field fortifications, sieges, and bridges. They stitched together their empire with hard-surface roads, some of them so well founded and paved that they have survived into the twenty-first century, roughly two millennia after their construction. What these roads all have in common with modern variants such as the German autobahn and the American interstate highway system is that they served civilian purposes of commerce and government administration while at the same time allowing states to mobilize and move their armies to sites of external threat. Unfortunately for many states through history, they also provided avenues of invasion, undermining their military purpose in the most disastrous way.

A more recent example of non-weapons dual-use military technology is the steam engine. A classic instance of technology push, the steam engine was first a scientific curiosity—appearing from ancient times through the seventeenth century—and then a commercial tool to pump groundwater out of coal mines. Those first steam engines of the eighteenth century were so inefficient that they made economic sense only when operating at the mouths of those coal mines, where fuel was cheap. Not until 1769, when James Watt invented the separate condenser, did the steam engine begin to realize its potential. Working with business partner and cannon manufacturer John Wilkinson, the company of Boulton and Watt provided the Wilkinson ironworks with power for its drills, while Wilkinson provided Boulton and Watt with a boring technology that made possible precision cylinders for their engines. It was a civil-military technological synergy seldom matched by any military-industrial complex. Soon steam engines were powering not only the factories of Britain's industrial revolution but also railroads to move armies in the American Civil War and the wars of German unification, and steam warships powerful enough to overcome wind and tide. Even the most modern major warships, fueled by oil or nuclear power, are steamships that drive themselves and their auxiliary equipment by passing steam through modern turbines.

Equally important in both warfare and civil society is the internal combustion engine. The first internal combustion engine was the cannon, an instrument for harnessing to human purposes the energy given off by rapid burning of a carbon compound in an enclosed space. The steam engine, an external combustion engine, used an external fire to heat water in a confined space. Not until the late nineteenth century did practical machines appear that transformed the energy of fire directly into mechanical power. Over the course of the nineteenth century, a series of experiments with internal combustion engines, accelerated by the commercial availability of petroleum distillates, produced practical machines operating on both spark and compression ignition.

By World War I, internal combustion engines were powering military aircraft, submarines, land vehicles for passengers and cargo, tanks, and even auxiliary electrical power. Wherever internal combustion engines and fuel could go, electricity could go, for lighting, heating, radios, telegraphs, machine shops, hospitals, kitchens, refrigerators, and the innumerable electrical appliances that support military operations. Huge, fixed generating plants had been powering cities since late in the nineteenth century. With the development of portable generators powered by internal combustion engines, armies could now campaign with all the appliances of modern warfare. And airplanes could carry aloft the radios, instruments, and auxiliary equipment that supported aircrews in their missions of fighting, bombing, and reconnoitering. The warfare of total war was empowered by the internal combustion engine.

Another non-weapons dual-use military technology that is also a mainstay of civil society is electric and electronic communication. This category includes everything from the first electric telegraph to the more modern telephone, radio, television, and Internet. These latest means of communication can carry analog and digital signals of codes, voices, and images at or near the speed of light. Of course smoke and flag signals had traveled at the speed of light throughout history, but they were limited to line of sight. (Sound certainly traveled at its own speed, but that was much slower than light.) Beginning in the nineteenth century, military personnel could communicate with each other near the speed of light if they were connected by powered wires; the introduction of radio sped the communication and eliminated the wires, though its range was constrained by a variety of technical and environmental factors.

In modern digital communications, all content is digitized—converted to binary form—and then transmitted on an appropriate electromagnetic wave, to be converted into data, voice, or visual form at the receiver. It travels at the speed of light in line of sight or broadcast, depending on the nature of the receiver. Because warfare has always been a zero-sum game, in which one side's advantage was the other side's disadvantage, the commander who learned of his enemy's actions or movements before his own were known to them, and who could deliver his orders to his subordinates faster than his adversary could, had an overwhelming battlefield edge. Today's military commander has at his disposal—for better or worse—almost instantaneous, worldwide communication of all forms of information up and down his chain of command and real-time contact with his subordinates in combat. It is a nice question whether this has cleared Clausewitz's "fog of war" or thickened it.

The first "computers" were women, civilians calculating ballistic trajectories for the army in the years leading up to World War II. During that war, computers, another non-weapons dual-use technology, became machines. The rise of these machines to their twenty-first-century ubiquity can be classed as neither military nor civilian. Both realms of society made indispensable contributions to the evolution of computers, no matter how one defines "computer." It is a commonplace of twenty-first-century life that a computer revolution has taken place—or is still taking place—but there is no consensus on what changed. Has it been a revolution in communications, information, computation, artificial intelligence, or simply entertainment? It may be best to think of it in technological terms as a refinement of solid-state electronic devices that has made possible significant transformations in all those fields of human activity.

From that perspective, the military made important contributions to early analog and digital computers for such purposes as ballistic firing tables, encryption and decryption of communications, simulation of nuclear reactions, and integration of radar networks. The first transistor emerged from civilian work on telephone switching, but one of the first two inventors of microprocessors, Jack Kilby, made his discovery while working for the US Air Force on the electronics of missiles. The military also played a critical role in the first networking of computers, and has made its share of contributions to subsequent developments as well. Now solid-state electronic devices of unimaginable complexity and capability empower military instruments ranging from the newest night-vision goggles to interceptor missiles that can achieve the mythical goal of hitting a bullet with a bullet. The so-called net-centric battlefield of the twenty-first century is awash in microcomputers and connected almost instantaneously to macrocomputers of superhuman calculating power. Ships, aircraft, spacecraft, and their payloads—including weapons—are systems capable of near-autonomous operation.

Isaac Newton illustrated his theory of gravity by hypothesizing an object propelled horizontally from a mountaintop in such a way that at some point the force of its movement tangential to the earth's atmosphere would exactly match the gravitational pull of earth. Such an object, he explained, would become a satellite of earth, balanced between escape velocity and the gravity well that is planet earth. It would take almost three centuries for humans to devise the launch vehicle necessary to test Newton's theory, but the military implications of the capability were obvious all along. Not only could a satellite—also an artifact of a non-weapons dual-use technology—observe the earth from space, but it could also de-orbit all or part of itself to strike a target on the earth's surface. The first human satellite, *Sputnik I*, entered orbit on October 4, 1957. Though its mission was nominally scientific, an experiment in the International Geophysical Year, its true impact was military. For, as Newton had explained, the force that could put a body in orbit also could fly it to a target on the other side of the world, where deceleration could bring it down upon a predetermined point. The launch vehicle was, by definition, an intercontinental ballistic missile. Defense intellectuals immediately prophesized the militarization—indeed the weaponization—of space, extrapolating from human experience on the sea and in the air that warfare would go wherever humans went.

As it turned out, the great powers have indeed militarized space, but so far they have by and large refrained from weaponizing it. So-called near-earth orbit, ranging from hardly more than 100 miles up to geostationary orbit more than 22,000 miles above the earth's surface, is awash with military satellites conducting communications, reconnaissance, signals interception, meteorology, and global positioning. In the Outer Space Treaty of 1967, the two superpowers, followed in subsequent years by most of the other nations of the world, forswore the placement of weapons of mass destruction in space. And the technology of satellites and their orbits has made it clear to most that conventional weapons circling the earth make no more sense than nuclear weapons. So satellites have become indispensable to military activities and operations on earth, but humans so far have seen fit to keep their weapons within the atmosphere.

This list of non-weapons dual-use technologies could be expanded easily. It might, for example, include canning of food, tracked vehicles, transport aircraft, helicopters, gyroscopes, radar, GPSs, and digital fly-by-wire for movement of aircraft control surfaces. But the principal significance and implication of these technologies remains the same. Beginning in prehistoric times, humans have developed non-weapons military technologies. Many were first developed for civilian purposes and then adapted to military functions; the Schöningen spears come to mind. But sometimes, as with computers and fortifications, the military played the leading role. In the modern world, some civilians are uncomfortable using technologies of war, even if they do not kill or destroy.

By the same token, militaries often believe that technologies developed for civilian purposes require extensive modification to make them adequate to the demands of warfare. More often, however, people are unaware of where these technologies came from and what purposes called them into existence. Few civilians worry that their automobile engines also power airplanes and submarines and tanks. Few e-mailers worry that their mode of communication evolved out of personal messages exchanged between researchers for the US Defense Advanced Research Projects Agency interacting on a network designed to share research results. Furthermore, the growing importance of non-weapons technologies in modern warfare demonstrates the accelerating trend toward conflicts spilling off the battlefield and into the civilian communities, transportation networks, economic markets, medical facilities, and industrial arenas of modern life.

Weapons dual-use technologies

Even weapons can be dual-use. Not all instruments of force in society are military. The state has—or claims—a monopoly of armed force within the territory it purports to control, but the state may license citizens outside the military to use force in certain prescribed circumstances, such as policing, self-protection, security, hunting, and the like. As with nonweapons technologies, these instruments may have military or civilian origins before migrating to the other realm. Also as with non-weapons technologies, a few examples will clarify the topic.

Pride of place among dual-use weapons goes to the Schöningen spear and its prehistoric cousin, the bow and arrow. From hunting animals to hunting people, these technologies proved equally effective in war and peace. Of course, the similarities between hunting and warfare included much more than technology. Both pursuits exploited intelligence, stealth, teamwork, communication, and courage, in addition to knowledge of terrain, weather, and behavior of the prey. Humans, in addition to being hunters, also could be the hunted, needing technologies of defense against both two-legged and four-legged predators. The primary tactic of the prehistoric hunt, as best we can surmise, was pounce and flee, the same tactic still used by relatively weak military forces against their stronger foes. We may think of it now as ambush, or what that great military strategist Mao Zedong called "mobile warfare," but the principles are the same. Use surprise to attack the prey unawares, but save an escape route to run away and fight another day if need be. Thus it was that the first dual-use weapons were missile weapons, inflicting wounds at a distance while allowing the attacker avenues of flight.

The chariot is another dual-use weapons technology that has already been discussed. It is important to note, however, that it always had both weapons and non-weapons functions in the military. In this sense, it was like ships, planes, rockets and other weapon platforms. More than a weapon, it was part of a weapon system, whose parts could be disaggregated into categories of weapons and platforms. Indeed, the chariot was a quintuple-use technology, dominating combat in the Levant through much of the second millennium BCE and then taking up noncombat roles such as transport, hunting, racing, and ceremony. As with ships, airplanes, and spacecraft, its military version combined a moving platform with an onboard weapon system. It was this basic configuration that made it a natural dual-use technology, for the platform always held out the promise of alternative uses, just as ships and planes can serve civilian functions. When the chariot was used as a jeep in its transportation role, carrying Achilles to his showdown with Hector, for example, it functioned the same as dragoons and modern infantry riding to battle in armored personnel carriers—almost, but not quite, a weapon system. But the chariot in its other roles was strictly civilian. No doubt many of the ceremonial uses of the chariot—Roman triumphs come to mind—sought to bestow an aura of military prowess on the returning champion, but this use was no more military than the Constantinople chariot races between competing political parties in the Byzantine Empire.

Nuclear power is another dual-use technology. This one gave the military both weapons (bombs) and non-weapons (ship propulsion) uses. A science-based technology, it arose from rapid advances in theoretical and experimental physics in the 1930s. It was physicists in the United States, some of them refugees from Nazi Germany, who first brought to the attention of President Franklin Roosevelt the possibility of an atomic bomb. The crash program of the Manhattan Project during World War II resulted in the only use of atomic weapons in warfare in human history. Nuclear weapons went on to have a profound impact on war, but only a secondary effect on warfare—for these weapons have never again been detonated in anger. Rather, they contributed to the "Long Peace," the absence of great-power war since 1945. Like fortification, the nuclear revolution has been as important for the wars that did not happen as for the ones that did. All warfare since 1945 has been shaped by the nuclear umbrella under which it has operated.

Meanwhile, peaceful uses of nuclear power have proliferated. These have been most prominent in the generation of electricity and in medicine. Attempts have been made to use nuclear propulsion for ships—and even airplanes—but the only widespread application has been in military submarines and capital ships—mostly American aircraft carriers. The first use of this technology to destroy the Japanese cities of Hiroshima and Nagasaki, along with the accidents, both civilian and military, that have occurred from time to time in the ensuing decades, have cast a pall of danger and fear over nuclear power. Yet Admiral Rickover and others demonstrated that it could be used safely when handled with care.

Equally ironic in its suspension between military and civilian uses is chemical weaponry. These wicked instruments of death and disability rose to prominence in World War I, when the German chemist Fritz Haber bent his Nobel Prize–winning talents to the development of chlorine gas and other deadly gases. Haber's postwar defense that death was death by any means ignored the horrendous suffering and disability inflicted by some chemical weapons, such as mustard gas. But Haber might have made a different argument for chemical weapons, as others have done: they could put soldiers out of action without killing them. Indeed, by this standard, mustard was a more humane agent than the deadlier chlorine and phosgene, because it was less lethal. But even if advocates of gas warfare had been able to overcome the world's moral revulsion at these weapons, they still faced intractable problems of delivery. Neither exploding shells nor canisters, let alone aerial bombs, could ensure that the released gas would not carry on the wind onto friendly forces or innocent civilians. Thus it was that the technology of distribution posed a greater challenge than the agents themselves, leading to post–World War I reaffirmations of the 1907 Geneva Convention against gas warfare. That taboo held over the ensuing century, with a few horrible and frightening exceptions—mostly against civilians.

The primary civilian analog of gas warfare is the release of chemicals that attack human pain receptors. Tear gas and pepper spray are the most common. Ironically, tear gas is classified as a chemical weapon by the Geneva Convention, and therefore banned from warfare. But most states use it against their own citizens for subduing criminals and controlling crowds. Both agents have the potential to kill, though exposure to them seldom results in death. Still, their continued use points to the blurring of distinctions between civil and military realms in the modern world.

Another dual-use weapons technology—explosives—may seem at first blush to be so exclusively military as to disqualify itself. But explosives appear to have originated in China as fireworks, and their civilian uses continue to shadow their more familiar combat roles. All conventional nonnuclear explosives share the same physical profile: they derive their power from a chemical reaction that takes the form of rapid and confined burning. Gunpowder, the first and most revolutionary of explosives, underwent constant research and development from the time of its introduction in the West by the Mongols in the thirteenth century. All varieties combined carbon, sulfur, and saltpeter (potassium nitrate). The trick was finding the right proportions, which varied depending on the purity of the ingredients. By the nineteenth century, researchers were exploring variations to achieve greater power, smaller bulk, and less smoke. The results included TNT, smokeless powder, gun cotton, nitroglycerin, dynamite, and various plastic explosives. The military sponsored much of the research behind these developments, but the results found countless civilian applications. Of course, fireworks continue to amuse and entertain, but explosives also aid mining, civil engineering, demolition, avalanche control, and other constructive pursuits. Military explosives also power the small arms used by hunters, sportsmen, and peace officers, and nitroglycerine ameliorates some heart conditions.

Missiles and rockets have varying, overlapping, and confusing definitions that invite misunderstanding. For purposes of this discussion, it is best to think of rockets as self-propelled projectiles driven by the rearward thrust of hot gases produced in the combustion of fuel and oxidizer carried within the vehicle.

A missile may be any projectile, but here the term will refer to those rockets that are actively guided in flight. Rockets have flown since the first Chinese fireworks, but the first applications of propulsive combustion in the West were not in rockets but in guns, where the propellant burns explosively and throws the projectiles without further application of force after leaving the gun. The first military rockets in the West appeared in the late eighteenth century and received some lasting fame in the attack on Fort McHenry in the War of 1812, when Francis Scott Key immortalized their red glare. But because early rockets were unguided, they remained area weapons of limited effectiveness until the middle of the twentieth century.

Then Wernher von Braun and his colleagues combined a rocket flying a ballistic trajectory with a crude inertial navigation system to direct their V-2s hundreds of miles and land them in a circular error probable (the circle within which 50 percent of the rockets could be expected to fall) of 4.5 kilometers. Guided missiles went on to become a cornerstone of the strategic arms race between the United States and the Soviet Union and even now are a guarantor of great-power peace. But those same rockets that maintained the balance of terror between the superpowers also served as the launch vehicles of the space age. Virtually all spacecraft that have left the earth's atmosphere since the flight of *Sputnik I* in 1957 have ridden on the technology of ballistic missiles, both solid- and liquid-fueled. The core technologies were developed by the military for military purposes. And Wernher von Braun again represents the military-civilian dynamic. He began in civilian pursuit of spaceflight, migrated to military work for the Wehrmacht and the US Army, and returned to civilian pursuits to build the Apollo launch vehicle to carry Americans to the moon. The von Braun paradigm still empowers and constrains human spaceflight.

The final dual-use technology in this compilation is automatic firearms, or machine guns. These instruments are individual or crew-served weapons that employ mechanisms for clearing the chamber, inserting a new round from a belt or magazine, and firing that round without any discrete input from the gunner beyond constant pressure on the trigger. Since the first gunner stepped onto a battlefield, his rate of fire has been a main determinant of success. Indeed, the first gunners took so long to reload that they had to be protected by pikemen lest enemy cavalry fall upon them between shots. A series of innovations from the seventeenth century to the twentieth increased rates of fire by replacing matchlocks with flintlocks, combining shot and explosive in a single cartridge, loading in the breech instead of the muzzle, and employing bullets with percussion caps and extractable shells, muscle-powered mechanisms to expel shells and insert new bullets and cock the weapon, and finally gas-powered reloading mechanisms that used the power of the bullet's explosion to perform the same functions. After that, it was just a matter of improved design to produce ever faster, lighter, and more reliable automatic weapons. Americans had a special knack for this line of development, perhaps because of their national preoccupation with the right to bear arms. Not only did Americans lead the development of the machine gun for military purposes, but they also led the world in introducing automatic weapons in hunting, sport, and personal security. At the time of this writing there are more personal firearms in America than there are Americans, many times more than the number maintained by the United States military. And most of these weapons were developed in the first instance for military purposes. It is difficult to think of a military technology that has permeated civil society more fully than the individual firearm in America.

What, then, might we say about dual-use technologies, both weapons and non-weapons? First of all, they illuminate the fundamental question of whether or not military research and development and production benefit society. Is there a redeeming civilian spin-off from research devoted to purposes of warfare? In some cases, there surely has been, though any such redemption must be discounted for the opportunity costs of what those researchers might have contributed to society had they worked on civilian technologies directly. In the same vein, have the economies of modern, industrialized states become dependent on government spending for military research and development? Have the world's major free-enterprise democracies become, in William McNeill's terms, command economies, channeling their resources into state purposes and starving the free market? Or have these democracies become "national security states," in the language of Michael Hogan and other historians? The military-industrial complex of the Cold War has loosened its grip on most developed states, but it has not disappeared. Furthermore, we might wonder if modern military technology, like war itself, is spreading throughout human societies, blurring the former distinctions between military and civilian, combatant and noncombatant, war and peace. If military technologies pervade civilian life, and if civilian technologies are appropriated to military purposes, then the militarization of the modern world may well run more deeply in the fabric of modern society than we are wont to admit. Dual-use technologies shed light on all these questions.

Military revolutions

As soldiers and scholars contemplated technological changes in warfare in the 1990s, two arcs of analysis intersected without really having much impact on each other, passing instead like proverbial ships in the night. But the similarities and differences in their trajectories speak volumes about our understanding of the technology of warfare at the beginning of the twenty-first century. They also highlight the risks inherent in thinking superficially about this topic. And they illuminate the ways in which military technology has evolved since World War II.

Military historians described one arc of analysis, the role of military revolutions in history. Historian Clifford Rogers has shown that the term "military revolution" had been a trope of military commentary and analysis in the West through the eighteenth and nineteenth centuries. But the term gained purchase on the historical imagination only when scholars began to think of it in the same way they thought about the momentous Western revolutions that altered the course of history—most prominently the American, French, Russian, scientific, and industrial revolutions. Historian Michael Roberts intimated just such a comparison in his 1955 lecture "The Military Revolution, 1560–1660." Roberts described a transformation of ground warfare in Europe prompted by the introduction of individual firearms and field artillery on the battlefield in the early modern period (roughly 1500–1789). The transformation was characterized by new tactics integrating firearms and pikes, large and sustained campaigns, bigger armies, and a greater impact of warfare on society. The historiography of Europe between the Renaissance and the French Revolution was then in what one historian has called the "early modern muddle," and Roberts's thesis added salience and gravitas to the discourse. It also highlighted the contributions of Swedish king Gustavus Adolphus (1594–1632), whose biography Roberts was writing.

Historian Geoffrey Parker endorsed Roberts's thesis in 1976, while revising it significantly. Then, in 1988, Parker completely reformulated the thesis in his landmark book, *The Military Revolution: Military Innovation and the Rise of the West, 1500–1800*. By this time, Roberts's thesis was hardly recognizable. The only component that remained intact was an increase in army size, but this Parker attributed to the introduction of the *trace italienne*, a new style of fortification developed as a counter technology to siege artillery. Parker added two entirely new components to the early modern military revolution: the extension of its temporal boundaries to cover the entire early modern period and the projection of European power overseas in the first great wave of Western imperialism. Expanded in this way, Parker said, the military revolution explained, at least in part, the rise of the West. This larger and more potent military revolution certainly bore comparison with the great political and material revolutions of the Western historical canon. Parker's argument had been anticipated by other historians, but his book nonetheless took the military history community by storm, becoming one of the two most influential works of the last fifty years, along with John Keegan's *The Face of Battle*. It set off a tsunami of scholarship criticizing the Parker model, finding other examples, and theorizing the phenomenon of military revolution. Since historians usually find what they go looking for, the literature began to fill up with histories of military revolutions. They were found in the Middle Ages, in Asia, in the American Civil War, in the naval arms race at the turn of the twentieth century, and in the wars of German unification, to name just a few instances. Almost all of these examples generated their own definitions of what constituted a military revolution, expanding and diluting the concept at the same time.

Meanwhile, another scholarly trajectory was rising out of a different intellectual community: American soldiers and defense analysts. It came to be called the "revolution in military affairs" (RMA). The American defense intellectuals had the concept from Soviet military analysts, who had theorized a "military-technical revolution" in the 1950s. At first the Soviets focused on the impact nuclear weapons might have on the conduct of conventional warfare. In the 1960s and 1970s, those concerns evolved into a related unease about the growing gap between Soviet and American conventional military technology. The post–World War II enthusiasm for technological innovation within the American armed services was producing rapid technological change with which the Soviet Union simply could not keep pace. In realms such as computers, high-performance aircraft, stealthy submarines, satellite reconnaissance, and many other cutting-edge "high" technologies, the United States seemed to be moving into a realm by itself, a first among equals that might soon achieve unassailable preeminence over Soviet—and all other—military forces.

As Americans read this Soviet literature, they developed a new appreciation for their own ascendancy. Did it not make sense to concentrate on this asymmetric advantage over the enemy? Was not the Soviet concern proof of the efficacy of American research and development?

Thus was born in the American defense community a campaign to feed and strengthen the "revolution in military affairs." Precision-guided munitions, a pet project since the disappointments of Vietnam, were achieving unprecedented accuracies. Talk abounded of an "electronic battlefield" of the future. Air power theorist John Boyd preached "OODA loops," a doctrine using American technological sophistication to allow its forces to observe, orient, decide, and attack faster than the enemy. Visionaries spoke of "net-centric warfare," in which electronically networked forces would reconnoiter, communicate, and coordinate on the battlefield faster than their foes.

The revolution in military affairs emerged in various forms, but all had certain characteristics. None of the interpretations of the RMA were about nuclear warfare, either strategic or tactical. They *were* about America's qualitative edge in conventional military technology, a hedge against Soviet/Russian numerical superiority in land forces in Europe. And they all predicted that the United States might reach an unassailable plateau of military capability on which it would be invincible to all, including the Soviets/Russians. The movement accelerated in the 1990s. The capability was demonstrated to the satisfaction of its advocates in the first Gulf War (1990–1991). Andrew Marshall, guru of the Pentagon's Office of Net Assessment, sponsored a formal study of the phenomenon. And the administration of President Bill Clinton (1993–2001) considered it a way to cut defense spending without reducing American security. Part of the irresistible allure of the revolution in military affairs was that it seemed to offer more bang for the buck.

The RMA also attracted the attention of former secretary of defense Donald Rumsfeld. When Rumsfeld returned to the post of defense secretary in the administration of George W. Bush in 2001, he announced two major goals: to field an operational ballistic missile defense system (which had been under development since it was announced by Ronald Reagan in 1983) and to use the RMA to reform the military. In Rumsfeld's view, the Pentagon—especially the army—remained wedded to a Cold War paradigm of conventional war, which would be fought against larger Russian forces on the plains of Europe. This mindset was captured in the army's devotion to a next generation of mobile field artillery. The *Crusader*, six years into development when Rumsfeld re-entered office, was a tracked, self-propelled, automatic-loading 155-millimeter gun able to throw a 100-pound projectile 14 miles. Weighing 43 tons and towing a 40-ton resupply vehicle for fuel and ammunition, the gun could be airlifted on C-5A and C-17 transport aircraft to any crisis scene with a 3,500-foot runway. But Rumsfeld believed the army was preparing to fight the last war against a Soviet empire that no longer existed. He wanted a lean, nimble army to fight small wars. He canceled the *Crusader* soon after the attacks of September 11, 2001, on New York and Washington.

Rumsfeld relied on the capabilities attributed to the revolution in military affairs to respond to September 11. In Afghanistan, where the attacks had been orchestrated, the American military put a handful of "boots on the ground" to direct air strikes against the al-Qaeda enemy and its Taliban hosts. In a matter of weeks, American firepower had driven al-Qaeda into Pakistan and the Taliban into hiding. Then the Bush administration turned its sights on Iraq. Ignoring the advice of his army chief of staff, Secretary Rumsfeld invaded Iraq with a preliminary airpower campaign of "shock and awe," supported by about 150,000 American troops on the ground. This juggernaut rolled over the army of Saddam Hussein (weakened by the 1990–1991 Gulf War), drove Hussein into hiding, and "liberated" the country to face the Sisyphean task of building a stable and just state in a chaotic corner of the world. When President Bush appeared on a US

9. An F/A-18F Super Hornet prepares to launch into the night from the deck of the aircraft carrier USS *Harry S Truman*. The system of systems embodied in such nuclear-powered carriers is the most complex military artifact so far in the twenty-first century.

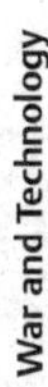

aircraft carrier in the Persian Gulf on May 1, 2003, under a banner reading "Mission Accomplished," it was a testament to the revolution in military affairs. The United States did indeed seem to have an irresistible military prowess.

Soon, however, American ground forces revealed their own Achilles heel. One by one, the vehicles of the new, modern, mounted warfare succumbed to ambush by IEDs—improvised explosive devices. These simple bombs, detonated by contact, timing, or command, soon infested the Iraqi roads and bridges on which American military vehicles traveled. The detonating instruments were as simple as a cell phone. The explosives ranged from hand grenades and small mortar and artillery rounds (many captured from Americans or from Iraqi army ammunition depots abandoned when the regime collapsed) to massive unexploded bombs and homemade charges. US trucks, armored personnel carriers, and even tanks were unprepared for this weaponry. They

10. The improvised explosive device (IED) is the ultimate counter technology. This ordinance was captured by coalition forces in Baghdad during the war in Iraq (2003–2013). Mines and artillery shells such as these were hooked to simple detonators, such as cell phones, and planted in the path of coalition forces.

were put out of action at alarming rates, and their crews and passengers were subjected to horrendous physical and psychological injuries. It would be years before the revolution in military affairs produced counter technologies equal to the challenge. Al-Qaeda even distributed online propaganda films of their IED attacks on the Americans, yet another dual-use technology turned against the industrialized West.

This was not the first time that high-tech, industrialized, Western-style armies had met setbacks imposed by low-tech, preindustrial, non-Western partisans. Mao Zedong had introduced what he called "people's war" in the civil war for China against the Western-style army of Chiang Kai-shek. Ambush, which he called "mobile warfare," loomed large in his scheme. His tactics were used by Ho Chi Minh's forces in Vietnam to capture the French garrison at Dien Bien Phu and then to defeat the American military in the final phase of the Vietnamese war of national liberation. Other wars in the years since World War II have pitted poorly armed partisans against the military establishments of industrialized states with equally surprising results. For example, Israel has twice struggled to win the battle against intifadas without losing the war of world public opinion. The terrorists who attacked the United States in 2001 inflicted more casualties on American soil than the Japanese attack on Pearl Harbor in 1941 using no weapon more sophisticated than a box cutter. Ironically, the weapon of choice for these partisans and terrorists has usually been the very weapon that marginalized the barbarians since the introduction of gunpowder: gunpowder.

What, then might be said about the state of military revolution in the early twenty-first century? First and most important is a caution: belief in technology—even high technology—as a military panacea is misplaced and dangerous. Technology does indeed favor victory, but it does not guarantee it. In studying revolutionary military technology, historians did better than the defense analysts, in part because the historians were retrospectively analytical and the RMA was prospectively prescriptive. While historians have shown themselves to be just as cavalier when invoking revolutions, they have at least had the insight to understand that revolutions can be identified only after the fact. Not all change turns out to be fast enough and great enough to warrant the label "revolutionary."

Furthermore, the RMA and the trope of "military revolutions" both involved a certain amount of professional gamesmanship. Historians could have more impact on the existing scholarship and sell more books by claiming that their studies revealed revolutionary change. And advocates of the RMA could exert more influence on policymakers by promising transformational change at low cost. This is not to say that claims of revolution were disingenuous, only that the rhetoric of revolution often proved irresistible to advocates and audience alike. Both experiences suggest that talk of revolution should always be received skeptically.

The final point of comparison between the two movements confirms all of these trends. The RMA group often invoked the historical literature on military revolutions because it seemed to add scholarly gravitas to the current phenomenon they were espousing. But students of historic military revolutions paid little attention to the discourse on the revolution in military affairs. The historians, after all, were mostly academics, swimming in a scholarly sea of left-leaning, antimilitary sentiment. Indeed, for reasons of theoretical and scholarly integrity, it behooved them to eschew practical, contemporary applications of their findings. Enthusiasm for the RMA and for military revolutions rose to prominence in the 1990s and 2000s, before fading in the 2010s.

Conclusion

Only a few words can or need be said about the future of technology and warfare. The increasing pace of technological change became a cliché of the twentieth century, and so it remains. But the kernel of truth behind the platitude is that the pace really is accelerating and is likely to continue doing so. This is especially true for military technology, which is still subject to widespread, self-conscious, institutionalized research and development. On the horizon as this book goes to press are true drones (not remotely piloted), robotic (preprogrammed) weapon systems, further microminiaturization, nanotechnologies of warfare, and, perhaps most alarming of all, autonomous weapon systems (capable of some degree of independent response to environmental and situational inputs). These will enter a world that is paradoxically more dangerous and less lethal than at any time in human history. That is, the technologies of warfare are more effective than ever before, but there is less warfare in the world, based on casualties as a percentage of population, than ever before. Where all this will lead is impossible to predict.

If this book suggests any answers, they probably lie in an understanding of the terms in the glossary. Military technologies will surely change in the future, but the principles behind these terms, like the venerable "principles of war," will probably abide. They appear to transcend peculiarities of time and place. No doubt other principles from other realms of human activity will also shape the future of warfare, but the glossary nonetheless provides a beginner's set of tools—a primer for our reborn Alexander—for thinking about the very particular realm of technology and warfare. The heavy emphasis in this book on early warfare makes the point that the concepts guiding the evolution of technology and warfare emerged early and remain potent.

Dual-use technologies, for example, have thrived across all of human experience, from the Schöningen spear to remotely piloted aircraft. It is reasonable to expect that civilian technologies will continue to find military applications and vice versa. And we can expect that attempts to limit the transfer of military technologies will be confounded by the dual natures of some. Cold War constraints on export of computer technologies, for example, proved difficult to enforce. Dual-use technologies also illuminate another truism of world history, that military power mirrors economic power. So true is this that economic competition in the world is increasingly seen as a kind of moral equivalent of military might.

So long as the world remains divided between developed states and those without industrial infrastructure, armed conflict between the two divisions will be predominantly asymmetric. Beyond this, it is impossible to predict what technological marvels the developed states will bring to bear or what low-tech innovations—IEDs, sabotage, purloined weapons of mass destruction, etc.—the have-nots will deploy. Symmetrical arsenals among the developed states will likely deter interstate war indefinitely, barring some technological breakthrough.

Cyber warfare offers one example currently gripping the public imagination. It seems at first to pose an unprecedented threat to developed states with complex networked infrastructures, who find themselves vulnerable to hackers—the new barbarians at the gates—with stealthy and irresistible siege technology at their disposal. Some of the concepts developed in this book can help to demystify the phenomenon and set it in historical context. First of all, cyber attacks so far have been for normative purposes of espionage, sabotage, and subversion—not warfare. Even the most serious cyber attack to date, the Stuxnet incursion into Iran's nuclear program in 2009 and 2010, failed to produce war. Cyber attacks constitute a dual-use technology that can target both military and civilian targets. They can be both symmetrical (e.g., between state actors) and asymmetrical (between state and nonstate actors). Cyber attacks operate in the tradition of missile weapons, working at a distance and allowing the attacker to escape direct retaliation; this recommends them to weaker adversaries, but states with superior cyber resources also have superior offensive potential. It was reportedly the United States and Israel that attacked Iran with Stuxnet. And, as with satellites, the gigantism of the Internet is a source of its vulnerability. North Korea reportedly escaped attack by a pre-Stuxnet virus by insulating most of its national computers from the Internet. All of this suggests that cyber attacks are simply a new form of technologies that the world has dealt with for millennia. No doubt cyber warfare will play a role in future conflicts, but the powerful state actors will have at their disposal superior resources, both to protect themselves and to retaliate against abusers of the system. Cyber warfare may well end up in the same category as poison gas and antisatellite weapons, in which the most powerful states will abstain from attacking each other and weaker states will attack to little effect.

The asymmetry of modern warfare seems to be producing a reversal of the classic preferences for missile and shock weapons. Historically, weaker powers have chosen missile weapons to ambush stronger enemies, while powerful states have tried to close with their weaker enemies and crush them. While many weaker combatants continue to use missile weapons to pounce and flee, those embracing suicidal warfare seem to be turning increasingly to shock attack—closing with the enemy to kill all within reach, including themselves. The world has seen suicidal warriors before—the Japanese kamikazes of World War II come quickly to mind—and the tactic has yet to prove sustainable. One reason, of course, is that this method uses up one's reserves of manpower. But technology alone cannot reveal whether the current instance will be different. Meanwhile, developed states find themselves turning increasingly to missile weapons, the old favorites of the barbarians. In modern warfare, these instruments are now called "standoff weapons," the drones and other tools of air power that avoid the risks of putting "boots on the ground." As with the automated weapon systems now being developed for future warfare, these attempts to win battles without putting military personnel in harm's way are without precedent in human history.

Dueling technologies will no doubt continue to evolve, so long as one side or another fields new technologies that are perceived by their adversaries as threats. The recent contest between IEDs and armored vehicles in Iraq and Afghanistan provides one example, as does the continuing refinement of ballistic missiles and antiballistic missile systems. In the latter case, however, if any community wanted to nuke a US city, for example, it would likely opt for a low-tech delivery platform: a ship in New York's East River or a container moving through the Los Angeles/Long Beach Seaport. In 2014, these two West Coast ports handled about 40 percent of the cargo entering the United States, on the order of seven million containers, each one a potential bomb carrier. A low-tech Trojan Horse can still be a better choice than a high-tech siege engine.

In the same vein, gigantism will probably continue the slow decline it has experienced in the nuclear age. Technological determinism will remain an empty epithet. Humans will retain the agency to harness their military technologies until human nature itself changes. And military revolutions, to say nothing of revolutions in military affairs, will remain rare. If Alexander returns, he will have a lot to learn, but the concepts explored here might provide a starting point.

Glossary

ambush (also **pounce and flee**): A tactic, often employed by the weak against the strong, using missile weapons. The attacker, often in a group, surprises the prey and inflicts as much damage as possible without getting in harm's way, then retires before the enemy can respond or be rescued by reinforcements.

appropriate technology: Few technologies are universal. To be successful, most must be appropriate, that is, suited to the time, environment, conditions, and applications in which they are applied. Galleys, for example, worked well in coastal waters but could not venture far out to sea.

asymmetrical technologies: A situational condition in which two sides engage in armed conflict with significantly different instruments of warfare, both weapons and non-weapons technologies. Since World War II, for example, aircraft carriers have enjoyed an asymmetrical advantage over conventional capital ships: the ability to attack them before coming within range of their guns.

capability greed: Historian Blair Hayworth's name for the propensity of military organizations to gold-plate their arms and equipment and to add unnecessary features.

Carbon Age: The second age measured by sources of military power. It ran from roughly 1400 to 1945, in between an age of muscle and wind power and the age of nuclear power. In the **Carbon Age**, firepower and machines driven by internal combustion engines dominated all realms of warfare.

cavalry-infantry cycle: Alternating dominance of mounted and infantry forces in land warfare.

closure (see also **lock-in, momentum**): A term from the social studies of science and technology identifying the point at which one of several possible technological pathways achieves such dominance of the marketplace as to virtually extinguish competition.

Combined-Arms Paradigms: Periods of land warfare in which all combatants fought with the same combinations of types of weapons, even though individual weapons and fighting styles varied greatly from state to state. After the chariot revolution, field warfare was conducted by combining mounted warriors and infantry. After the gunpowder revolution, field artillery added a third arm to the paradigm.

counter technology: A military technology designed to negate or reverse the effect of another technology.

demand pull (see also **technology push**): Technological development impelled by demand for some capability. Necessity is the mother of invention.

dual-use: Those technologies with both military and civilian applications.

dueling technologies: Technologies developed dialectically in response to each other's evolving capabilities. One example is fortifications versus siege technologies.

gigantism: Increasing the size or power of a technology in the belief that more is better.

lock-in: A term from economics marking the point at which producers of a commodity have invested so much (sunk costs) in a technological choice that it is considered impractical to backtrack to another path. See also **closure** and **momentum**.

military revolution: A transformation of warfare so profound and sweeping that it not only redefines armed conflict between states but also changes the course of history, shifting the relationship between states and access to coercive power. This book identifies three: chariot, gunpowder, and atomic/nuclear weapons.

missile weapons (see also **shock weapons**): Weapons that strike from a distance without requiring contact with the enemy. Also known as "standoff weapons."

momentum: Historian Thomas P. Hughes's alternative to **technological determinism**. It allows that some technologies acquire permanence over time as infrastructure adapts to them in their present form, making it difficult for human agency to change the technological paradigm. The United States' embrace of light water nuclear reactors is an example.

non-weapons technologies: Military technologies that support warfare without directly attacking people or things.

path dependence: A technology is **path-dependent** when its mature form is shaped significantly by the course of its development. To be **path-independent** suggests that there is one best technological solution to a problem and that it would be realized no matter what course the development process took. This flirts with **technological determinism**.

pounce and flee: See **ambush**.

revolution in military affairs: An American military theory of the 1990s and 2000s. It maintains that improvements in US conventional military technologies, especially high technologies such as computers and computer networking, would give the United States unassailable dominance of the battlefield. It fell out of favor in the 2010s.

shock weapons (see also **missile weapons**): Those weapons, such as sword, pike, and bayonet, that required closing with the enemy. At sea, ramming and boarding are **shock** tactics.

symmetrical technologies: Weapons and non-weapons military technologies that mirror the enemy's.

system of systems: Multiple technologies or artifacts gathered in an integrated combination with capabilities greater than those of the components. The most basic steamship, for example, requires a steam generator, a machine to convert heat to mechanical energy, and a propeller of some sort to turn mechanical energy into propulsion.

technological ceiling: A limit imposed on a technology or system by the inadequacies of one or more components. True submarines were impossible before nuclear power.

technological determinism: A rhetorical label used in two ways. First, it suggests that technology independently and decisively controls historical outcomes. Second, it can suggest that technology is **path-independent**, following an inexorable course of development to some single, best configuration.

technological stasis: A condition of static technological development with little significant innovation.

technology push (see **demand pull**): This occurs when a technological capability spurs development of applications. The availability of steam propulsion, for example, transformed naval vessels.

weapon platform: A vehicle carrying a weapon or weapon system. Chariots, tanks, ships, airplanes, and spacecraft can all be weapon platforms.

weapon system: A technology of attack or defense that consists of several component technologies or technological artifacts. All **weapon platforms**, for example, are weapon systems, as was the mounted knight and mobile field artillery.

Further reading

Adas, Michael. *Machines as the Measure of Men: Science, Technology, and Ideologies of Western Dominance*. Ithaca, NY: Cornell University Press, 1989.

Anderson, J. K. *Hunting in the Ancient World*. Berkeley: University of California Press, 1985.

Basalla, George. *The Evolution of Technology*. New York: Cambridge University Press, 1988.

Black, Jeremy. *A Military Revolution? Military Change and European Society, 1550–1800*. Atlantic Highlands, NJ: Humanities Press International, 1991.

Chase, Kenneth. *Firearms: A Global History to 1700*. Cambridge, UK: Cambridge University Press, 2003.

Cipolla, Carlo. *Guns, Sails and Empires: Technological Innovation and the Early Phases of European Expansion, 1400–1700*. New York: Minerva, 1965.

Davis, R. H. C. *The Medieval Warhorse: Origin, Development and Redevelopment*. London: Thames & Hudson, 1989.

Drews, Robert. *The End of the Bronze Age: Changes in Warfare and the Catastrophe ca. 1200 B.C.* Princeton, NJ: Princeton University Press, 1993.

Edgerton, David. *The Warfare State: Britain, 1920–1970*. Cambridge, UK: Cambridge University Press, 2006.

Ellis, John. *The Social History of the Machine Gun*. New York: Pantheon, 1975.

Engels, Donald W. *Alexander the Great and the Logistics of the Macedonian Army*. Berkeley: University of California Press, 1978.

Ferrill, Arther. *The Origins of War: From the Stone Age to Alexander the Great*. London: Thames & Hudson, 1985.

Gaddis, John Lewis. *The Long Peace: Inquiries into the History of the Cold War*. New York: Oxford University Press, 1987.

Gat, Azar. *War in Human Civilization*. New York: Oxford University Press, 2006.

Hacker, Barton C., with the assistance of Margaret Vining. *American Military Technology: The Life Story of a Technology*. Westport, CT: Greenwood, 2006.

Hale, John R. *Lords of the Sea: The Epic Story of the Athenian Navy and the Birth of Democracy*. New York: Viking, 2009.

Hall, Bert S. *Weapons and Warfare in Renaissance Europe*. Baltimore: Johns Hopkins University Press, 1997.

Hanson, Victor Davis. *The Western Way of War: Infantry Battle in Classical Greece*. 2nd ed. Berkeley: University of California Press, 2000.

Haworth, Blair. *The Bradley and How It Got That Way: Technology, Institutions, and the Problem of Mechanized Infantry in the United States Army*. Westport, CT: Greenwood, 1999.

Headrick, Daniel R. *Power over People: Technology, Environments, and Western Imperialism, 1400 to the Present*. Princeton, NJ: Princeton University Press, 2010.

Headrick, Daniel R. *The Tools of Empire: Technology and European Imperialism in the Nineteenth Century*. New York: Oxford University Press, 1981.

Heather, Peter. *Empires and Barbarians*. London: Macmillan, 2009.

Hogan, Michael. *A Cross of Iron: Harry S. Truman and the Origins of the National Security State, 1945–1954*. New York: Cambridge University Press, 1998.

Holley, I. B. *Ideas and Weapons: Exploitation of the Aerial Weapon by the United States during World War I; A Study in the Relationship of Technological Advance, Military Doctrine, and the Development of Weapons*. New Haven, CT: Yale University Press, 1953.

Jomini, Baron Antoine-Henri. *Treatise on Grand Military Operations*. Translated by S. B. Holabird. New York: Van Nostrand, 1865, 1:252.

Keeley, Lawrence H. *War before Civilization: The Myth of the Peaceful Savage*. New York: Oxford University Press, 1996.

Kern, Paul Bentley. *Ancient Siege Warfare*. Bloomington: Indiana University Press, 1999.

Landels, J. G. *Engineering in the Ancient World*. 2nd ed. Berkeley: University of California Press, 2000.

Landers, John. *The Field and the Forge: Population, Production, and Power in the Pre-Industrial West*. New York: Oxford University Press, 2003.

Lee, Wayne E. *Barbarians and Brothers: Anglo-American Warfare, 1500–1865*. New York: Oxford, 2011.

Lee, Wayne E. *Waging War: Conflict, Culture, and Innovation in World History*. New York: Oxford University Press, 2016.

Lynn, John A., ed. *Feeding Mars: Logistics in Western Warfare from the Middle Ages to the Present*. Boulder, CO: Westview, 1993.

Lynn, John A., ed. *Tools of War: Instruments, Ideas, and Institutions of Warfare, 1445–1871*. Urbana: University of Illinois Press, 1990.

Mao Zedong. *On the Protracted War*. Beijing: Foreign Languages Press, 1954.

McDougall, Walter A. ... *The Heavens and the Earth: A Political History of the Space Age*. New York: Basic Books, 1985.

McNeill, William H. *The Pursuit of Power: Technology, Armed Force, and Society since A.D. 1000*. Chicago: University of Chicago Press, 1982.

McNeill, William H. *The Rise of the West: A History of the Human Community*. Chicago: University of Chicago Press, 1963.

Millis, Walter. *Arms and Men: A Study in American Military History*. New York: Putnam, 1956.

O'Connell, Robert L. *Of Arms and Men: A History of War, Weapons, and Aggression*. New York: Oxford University Press, 1989.

O'Connell, Robert L. *Soul of the Sword: An Illustrated History of Weaponry and Warfare from Prehistory to the Present*. New York: Free Press, 2002.

Parker, Geoffrey. *The Military Revolution: Military Innovation and the Rise of the West, 1500–1800*. 2nd ed. Cambridge, UK: Cambridge University Press, 1996.

Piggott, Stuart. *Wagon, Chariot, and Carriage: Symbol and Status in the History of Transport*. New York: Thames & Hudson, 1992.

Pinker, Stephen. *The Better Angels of Our Nature: Why Violence Has Declined*. New York: Penguin, 2011.

Polybius. *The Histories of Polybius*. Translated by Evelyn S. Shuckburgh. London: Macmillan, 1889.

Roberts, Michael. *Essays in Swedish History*. London: Weidenfeld & Nicolson, 1967.

Rogers, Clifford. "The Idea of Military Revolutions in Eighteenth and Nineteenth Century Texts." *Revista de História das Ideias* 30 (2009): 395–415.

Rogers, Clifford., ed. *The Military Revolution Debate: Readings on the Transformation of Early Modern Europe*. Boulder, CO: Westview, 1995.

Rogers, Clifford. "The Military Revolutions of the Hundred Years' War." *Journal of Military History* 57 (April 1993): 241–78.

Rogers, Will. *New York Times*, 23 Dec. 1929.

Roland, Alex. *The Military-Industrial Complex*. Washington, DC: American Historical Association, 2001.

Roland, Alex. *Underwater Warfare in the Age of Sail*. Bloomington: Indiana University Press, 1978.

Smith, Merritt Roe, ed. *Military Enterprise and Technological Change: Perspectives on the American Experience*. Cambridge, MA: MIT Press, 1985.

van Creveld, Martin. *Supplying War: Logistics from Wallenstein to Patton*. Cambridge, UK: Cambridge University Press, 1977.

van Creveld, Martin. *Technology and War: From 2000 B.C. to the Present*. New York: Free Press, 1989.

Waldron, Arthur. *The Great Wall of China: From History to Myth*. Cambridge, UK: Cambridge University Press, 1989.

White, Lynn, Jr. *Medieval Technology and Social Change*. New York: Oxford University Press, 1962.

Whitehead, Alfred North. *Science and the Modern World*. New York: Macmillan, 1941.

Winner, Langdon. *Autonomous Technology: Technics-Out-of-Control as a Theme in Political Thought*. Cambridge, MA: MIT Press, 1977.

“牛津通识读本”已出书目

古典哲学的趣味
人生的意义
文学理论入门
大众经济学
历史之源
设计，无处不在
生活中的心理学
政治的历史与边界
哲学的思与惑
资本主义
美国总统制
海德格尔
我们时代的伦理学
卡夫卡是谁
考古学的过去与未来
天文学简史
社会学的意识
康德
尼采
亚里士多德的世界
西方艺术新论
全球化面面观
简明逻辑学
法哲学：价值与事实
政治哲学与幸福根基
选择理论
后殖民主义与世界格局
福柯
缤纷的语言学
达达和超现实主义
佛学概论
维特根斯坦与哲学
科学哲学
印度哲学祛魅
克尔凯郭尔
科学革命
广告
数学
叔本华
笛卡尔
基督教神学
犹太人与犹太教
现代日本
罗兰·巴特
马基雅维里
全球经济史
进化
性存在
量子理论
牛顿新传
国际移民
哈贝马斯
医学伦理
黑格尔
地球
记忆
法律
中国文学
托克维尔
休谟
分子
法国大革命
民族主义
科幻作品
罗素
美国政党与选举
美国最高法院
纪录片
大萧条与罗斯福新政
领导力
无神论
罗马共和国
美国国会
民主
英格兰文学
现代主义
网络
自闭症
德里达
浪漫主义
批判理论

德国文学
戏剧
腐败
医事法
癌症
植物
法语文学
微观经济学
湖泊
拜占庭
司法心理学
发展
农业
特洛伊战争
巴比伦尼亚
河流
战争与技术

儿童心理学
时装
现代拉丁美洲文学
卢梭
隐私
电影音乐
抑郁症
传染病
希腊化时代
知识
环境伦理学
美国革命
元素周期表
人口学
社会心理学
动物

电影
俄罗斯文学
古典文学
大数据
洛克
幸福
免疫系统
银行学
景观设计学
神圣罗马帝国
大流行病
亚历山大大帝
气候
第二次世界大战
中世纪
工业革命